ION
CHROMATOGRAPHIC ANALYSIS
OF ENVIRONMENTAL POLLUTANTS

Edited by

EUGENE SAWICKI Chief
J. D. MULIK Research Chemist
E. WITTGENSTEIN Research Chemist
Sampling and Analysis Methods Branch
Environmental Research Center
Environmental Protection Agency
Research Triangle Park, North Carolina

ANN ARBOR SCIENCE
PUBLISHERS INC
P.O. BOX 1425 • ANN ARBOR, MICH. 48106

Library of Congress Catalog Card No. 77-92589
ISBN 0-250-40211-4

PREFACE

Methods for the analysis of anions and cations usually have been complicated, difficult, expensive and/or fraught with uncertainties. Ion chromatography was recently introduced by Small and his co-workers at Dow and is now filling this long-felt need for a reliable simple polypollutant method of analysis for the water-soluble ions. If one considers the first stage of this technique as the discovery and development of the basic instrumentation, this book could be considered as part of the second stage, *i.e.*, the application of this instrumentation to the analysis of inorganic anions and cations. As seen from today's perspective the third stage would probably be a four-pronged expansion of the technique into (1) a further development of the second stage, (2) the analysis of organic anions and cations, (3) the analysis of essentially neutral compounds through their derived anions and cations, and (4) a further development of the technique (*e.g.*, the instrument, columns, detector, etc.) which would include a speeded-up automated system of analysis for these various compounds.

Two large families of samples could be analyzed by this rapidly evolving technique; these include a wide variety of environmental and biotic samples. We define the various environments by the routes of entry into the human body of the chemicals in that environment, *e.g.*, air pollutants, inhalation; water pollutants, oral; cosmetics, dermal; food additives and pollutants, oral; drugs, parenteral, oral, inhalation; etc.

The biotic samples could be subdivided into cells, simple tissues, organs, biological fluids, urine and other excreta, and exhaled air. These samples would contain a large variety of biochemicals.

Many of these chemicals in the environmental or biotic samples would be anions or cations or could be converted to anions or cations.

Consequently, even with the instrument as presently constituted many of these chemicals could be assayed. To ensure healthy individuals and environments the key chemicals and the toxic and antitoxic (prevention) pathways in man and his environment would have to be continually investigated, updated and monitored.

In summary, we think this technique will see a wide variety of uses, many unthought of as yet, once even more of our ingenious analysts start utilizing the instrumentation.

E. Sawicki

To the individual who does the best (s)he can
with what (s)he's got.

CONTENTS

v

POTENTIAL OF ION CHROMATOGRAPHY

E. Sawicki

Sampling and Analysis Methods Branch
Environmental Sciences Research Laboratory
U.S. Environmental Protection Agency
Research Triangle Park, North Carolina

ABSTRACT

The potential of the ion chromatograph for the analysis of a
wide variety of anions, cations and their precursors in the environment and in biological mixtures is discussed.

INTRODUCTION

Ion chromatography is an analytical technique about which we will hear
much more in the future. It is a form of ion exchange chromatography
utilizing a separation column, a background ion suppressor column, and
various eluents in a novel procedure in which conductance can be used as
the mode of detection. This method, which uses the speed of liquid chromatography and has so much potential in the fields of inorganic and organic analysis, was developed by a research group at the Dow Chemical
Co.[1,2] This and other aspects of this elegant technique are discussed in
Chapter 2.

Mulik et al.[3] applied the technique to the analysis of sulfate and nitrate
in air pollution samples. The method is being further developed by groups
studying air pollution and other environmental problems.

A large number of methods for sulfate, nitrate and other anions are
available and are used in analytical studies on air pollution. The bewildering number of sulfate methods that can be found in the literature are

shown in Table 1.1. Most of these methods are fairly complicated and difficult to perform. We believe that the ion chromatographic method is superior to these methods and will eventually supplant most of them in the analysis of a wide variety of mixtures.

Table 1.1   Methods of Analysis for Sulfate

A. Precipitation Methods
   1. Gravimetry
   2. Turbidimetry and nephelometry
   3. Colorimetry
      a. Direct (measure color increase)
      b. Indirect (measure color decrease)
   4. Fluorimetry
      a. Direct
      b. Indirect
   5. Ring oven, visual
   6. Polarography
   7. Atomic absorption
   8. Ion selective electrodes
   9. Titrimetry
 10. Radiometry

B. Direct Analysis of Solutions
   1. Ion chromatography-conductimetry
   2. Ion chromatography-square wave polarography
   3. Ion selective electrodes
      a. Direct
      b. Indirect

C. Direct Nondestructive Particulate Methods
   1. X-Ray fluorescence
   2. Electron spectroscopy chemical analysis (ESCA)

D. Reduction Methods
   1. Colorimetry
   2. Fluorimetry
   3. Gas chromatography
   4. Coulometric polarography
   5. Oscillographic polarography

E. Thermal Methods
   1. Flame photometry
   2. High resolution mass spectrometry

In addition there is a bewildering variety of methods for nitrate. Some of these are shown in Table 1.2. Here again, we feel that the ion chromatographic method is superior in terms of sensitivity, selectivity, simplicity and precision.

Table 1.2   Available Methods for Nitrate

A. Nitration Methods
    1. Colorimetry
    2. Fluorimetry
    3. Gas-liquid chromatography
    4. High-pressure liquid chromatography
B. Reduction to Nitrite and Assay
    1. Colorimetry
    2. Fluorimetry
    3. Gas-liquid chromatography
C. Reduction to Ammonia and Assay
    1. Colorimetry
    2. Gas chromatography
D. Thermal Decomposition
    1. Chemiluminescence
    2. Mass spectrometry
E. Enzymic
    1. Colorimetry
    2. Fluorimetry
F. Chemical Decomposition
    1. Manometry
    2. Chemiluminescence
    3. Gas chromatography
G. Oxidation Methods
    1. Colorimetry
    2. Fluorimetry
H. Complex Formation and Extraction into Organic Solvents
    1. Colorimetry
    2. Atomic absorption spectrometry
I. Direct
    1. Electron spectroscopy chemical analysis
    2. Spectrophotometry
    3. Ion selective electrode
    4. Ion chromatography-conductimetry

We feel that with the help of the ion chromatographic technique, a reliable sampling method for atmospheric nitrate and sulfate could now be developed with little difficulty; this method would have negligible artifact formation, *i.e.*, of sulfate from sulfur dioxide and nitrate from nitrogen dioxide.

Carrying this idea a little further, precursors of sulfate and nitrate could be determined with the help of ion chromatography. An example of this application is the determination of sulfur dioxide. The sulfur dioxide could also be determined through sulfate following its collection on an appropriate filter paper or a resin, in the same way nitrogen dioxide could be determined as nitrite or nitrate.

Similarly, other anions and their precursors could be determined through ion chromatography. Examples of some anions are nitrite, sulfite, thiosulfate, iodate, bromate, cyanide, thiocyanate, disulfide, chromate, phosphate, carbonate and the various aliphatic and aromatic carboxylates and sulfonates. Thus, the acid fractions of airborne particles should be amenable to ion chromatographic analysis, as should the acid fractions of other environmental and biological mixtures.

GENOTOXIC ANIONS

Many of the genotoxic anions of possible importance in mutagenesis, carcinogenesis and teratogenesis could probably be advantageously analyzed through ion chromatography. Some of these anions, their genotoxicity and techniques used in their analysis are given in Table 1.3.[4-13]

Factory workers in contact with arsenate have shown increased rates of respiratory and skin cancer.[14] Sodium arsenate is mutagenic in a bacterial system[15] and teratogenic with Swiss-Webster mice.[16] The International Agency for Research Against Cancer has stated that "available studies point consistently to a causal relationship between skin cancer and heavy exposure to inorganic arsenic in drugs, in drinking water with a high arsenic content, or in the occupational environment."

An increased rate of respiratory cancer in employees of Dow Chemical and Allied Chemical who handle arsenic compounds has been reported.[17] Strangely enough, experimental investigations in laboratory animals have consistently failed to produce evidence of the carcinogenicity of arsenic. This fact would seem to indicate that arsenic could be a cocarcinogen and/or that other cofactors are involved in the carcinogenicity of arsenic. The increased lung cancer rates in counties with copper, lead, zinc smelting and refining industries, but not in counties where other nonferrous ores are processed, suggests the influence of community air pollution from industrial emissions containing inorganic arsenic with arsenic, which is believed to play the primary role as a carcinogen, and other industrial agents contributing to the hazard. A single intraperitoneal injection of sodium arsenate in Swiss-Webster mice resulted in a teratogenic effect, depending on the day of pregnancy the animals were injected. Trivalent arsenic is stated to be much more toxic than pentavalent, both acutely and chronically. Pentavalent arsenic is often found in metallo-arsenicals, and is of concern because it can degrade into the trivalent form.

The evidence indicates that chromate causes lung cancer in workers in chromate-producing industries,[18] chromosome aberrations in the bone

Table 1.3 Atmospheric Inorganic Anions—Analysis and Genotoxicity

Anion	Analysis[a]	Genotoxicity[b]
Arsenate	AA,[4] IC, NA,[5] SP,[6] SSMS[7]	C, CC, M, T
Arsenite	AA,[4] IC, NA,[5] SP,[6] SSMS[7]	C, CC, M, T
Chromate	AA,[4] IC, SP[9c,10d]	C, CL, M
Nitrate	IC[3]	PC, PM[e]
Nitrite	IC M, PC, PM	M, PC, PM
Selenate	AA, GC,[11f] NA,[12] SPF[13g]	AC, C?, Cl[h], T[i]
Selenite	AA, GC,[11f] NA,[12] SPF[13g]	AC, C?, Cl[h], T[i]
Sulfate	IC[3]	CC
Sulfite	IC	M[j]

[a] AA = atomic absorption, GC = gas chromatography, IC = ion chromatography, NA = neutron activation, SP = spectrophotometry, SPF = spectrophotofluorimetry and SSMS = spark source mass spectrometry.

[b] AC = anticarcinogenic, C = carcinogenic, CC = cocarcinogenic, Cl = clastogenic, M = mutagenic, PC = precarcinogen, PM = premutagen and T = teratogen.

[c] Monitor reaction rate of the catalytic (chromate) oxidation of o-dianisidine by hydrogen peroxide by following the absorbance at 450 mm as a function of time for 15 min.

[d] Reaction of chromate with s-diphenylcarbazide and measurement of derived chromogen at λ 540 mm.

[e] Condensates from cigarettes treated with magnesium nitrate contained frameshift mutagens (which did not require microsomal activation) and mutagens, which could cause base-pair substitution mutations.

[f] The analysis is based on chelating Se(IV) with 5-nitro-o-phenylenediamine to form the thermally stable and volatile 5-nitro-2,1,3-benzoselenadiazole, which is extracted into toluene, separated by GC and, finally, determined with a microwave emission spectrometric detector.

[g] Selenium was all converted to the dioxide, reacted with 2,3-diaminonaphthalene, and the resulting selenadiazole was determined fluorimetrically at F370/517.

[h] Chromosome aberrations in cultured human leukocytes.

[i] Selenium is stated to cause reproductive failure and peculiar malformations in chick and mammalian embryos.

[j] A mutagen that exerts its effect by directly modifying DNA bases in such a way that base-pair errors arise at subsequent cell divisions.

marrow of rats[19] and mutagenesis in bacteria.[20] Chromium compounds of valence 3 and 6 produce tumors in rats. IARC states that there is an excessive risk of lung cancer among workers in the chromate-producing industry. Potassium dichromate in chronic and acute poisoning causes a significant increase in frequency of cells with chromosome aberrations in bone marrow of rats. Nitrate and nitrite can be considered precarcinogens and premutagens since, in the presence of appropriate amines, they can form nitrosamines, and in some cases this reaction can result in carcinogenesis or mutagenesis.[21-24] Nitrous acid has also been shown to be mutagenic to *Salmonella typhimurium*.[25] The genotoxic properties of selenium (in terms of its anticarcinogenicity, carcinogenicity, clastogenicity and teratogenicity) are highly controversial.[26] In states with higher selenium levels, there was significantly lower mortality in both males and females for several types of cancers, particularly the environmental problem indicators such as gastrointestinal and urogenital types of cancer. It has been reported that dietary selenium in animals causes striking inhibitory effects against a variety of carcinogens and even carcinogenic malonaldehyde, a product of peroxidative tissue damage. Further work is needed to confirm or refute the carcinogenic and anticarcinogenic results obtained with selenium. This is all the more reason to develop methods of analysis for selenium dioxide, selenite and selenate. More thorough analytical studies should help solve this problem. The cocarcinogenic activity of sulfate is shown by the inhibition of the hepatocarcinogenicity of N-2-fluorenylacetamide when the sulfate pool is depleted.[27] The mutagenicity of sulfite has been described.[28] Much more genotoxic data is available for these anions. Consequently, their analytical study would certainly be worthwhile.

OTHER APPLICATIONS

Cations which can be detected by ion chromatography include the alkali metals, the alkaline earth metals, ammonium, alkyl amines, dialkylamines, trialkylamines and tetra-alkylammonium salts.

Carcinogens determinable as cations include aromatic amines (such as β-naphthylamine, benzidine, 4-aminobiphenyl, 4-aminostilbenes, 4-phenylazoanilines), hydrazines, aza arenes and aziridines.

Ion chromatography should definitely be tried in the analysis of the basic fractions isolated from atmospheric vapors, particulates and other environmental materials. Many other compounds could be determined through ion chromatography as listed in Table 1.4.

The technique should be applicable to a wide variety of organic compounds which could be converted to anions or cations through decomposition and additive and other synthetic organic reactions. Aldehydes could be determined by analyzing the adducts formed with sodium bisulfite, nitroso compounds could be reduced to ammonium cations, and polyhydroxy compounds (*e.g.,* carbohydrates) could be oxidized to formic and other carboxylic acids. Key biological chemicals, such as histamine, and many metabolites of carcinogens and drugs could probably also be determined through the ion chromatographic technique.

Table 1.4 Ion Chromatography

F^-, Cl^-, Br^-, CN^-, SCN^-	Li^+, Na^+, K^+, R_4N^+
NO_3^-, NO_2^-, SO_3^{--}, SO_4^{--}	Alkaloids
AsO_3^{---}, AsO_4^{---}, PO_4^{---}	Amines + Imines
SeO_3^{--}, SeO_4^{--}, RCOOH	Aliphatic, Aromatic
Inorganic Gases	Heterocyclic
NO_2, SO_2	Biological
Aldehydes	Histamine
CH_2O, $CH_2{=}CHO$, $CH_2(CHO)_2$	Putrescine
Amino Acids	Aza Arenes
AMP, ADP, ATP	Drugs, Dyes
Coenzymes	Hydrazines
Phenols	Nucleic Acids, Bases
Nucleotides	NADH, NADPH, NAD^+, $NADP^+$
Estrogens	Carcinogenic Metabolites

With the introduction of new types of resins, precolumns, detectors, automation, faster analysis times, and a wide variety of ion formation methods, ion chromatography should see wide application in the fields of environmental carcinogenesis research, air pollution, water pollution, solid analysis, toxicology, food and drug analysis, pesticide analysis, industrial quality control, clinical chemical assay of biological fluids, cosmetics analysis and biochemical assays of tissues and cellular components.

REFERENCES

1. Small, H., T. S. Stevens and W. C. Bauman. "Novel Ion Exchange Chromatographic Method Using Conductimetric Determination," *Anal. Chem.* 47(11):1801 (1975).
2. Small, H. and J. Solc. *Theory and Practice of Ion-Exchange* (Cambridge, England: International Conference, 1976).
3. Mulik, J., D. Puckett, D. Williams and E. Sawicki. "Ion Chromatographic Analysis of Sulfate and Nitrate in Ambient Aerosols," *Anal. Letters* 9(7):C53 (1976).
4. Walsh, P. R., R. A. Duce and J. Fasching. "Impregnated Filter Sampling System for Collection of Volatile Arsenic in the Atmosphere," *Environ. Sci. Technol.* 11:163 (1977).
5. Morrison, G. H. "Spark Source Mass Spectrometry for the Study of Geochemical Environment," *Ann. N. Y. Acad. Sci.* 199:162 (1972).
6. Thompson, R. J., G. B. Morgan and L. J. Purdue. "Analysis of Selected Elements in Atmospheric Particulate Matter by Atomic Absorption," *Atomic Absorption Newsletter* 9:53 (1970).

7. Tabor, E. C. *et al.* "Tentative Method of Analysis for Arsenic Content of Atmospheric Particulate," *Health Lab Sci.* 6:57 (1969).
8. Kneip, T. J. *et al.* "Tentative Method of Analysis for Chromium Content of Atmospheric Particulate Matter by Atomic Absorption Spectroscopy," *Health Lab Sci.* 10:357 (1973).
9. Kneebone, B. M. and H. Freiser. "Determination of Chromium(VI) in Industrial Atmospheres by a Catalytic Method," *Anal. Chem.* 47: 595 (1975).
10. Abell, M. T. and J. R. Carlberg. "Simple Reliable Method for Determination of Airborne Hexavalent Chromium," *Am. Ind. Hyg. Assoc. J.* 229 (1974).
11. Talmi, Y. and A. W. Andren. "Determination of Selenium in Envitonmental Samples Using Gas Chromatography With a Microwave Emission Spectrometric Detection System," *Anal. Chem.* 46:2122 (1974).
12. Pillay, K. K. S., C. C. Thomas, Jr. and J. A. Sondel. "Activation Analysis of Airborne Selenium as a Possible Indicator of Atmospheric Sulfur Pollutants," *Environ. Sci. Technol.* 5:74 (1971).
13. Johnson, H. "Determination of Selenium in Solid Waste," *Sci. Technol.* 4:850 (1970).
14. Hill, A. B. and E. L. Fanning. "Cancer from Inorganic Arsenic Compounds. I. Mortality Experience in the Factory," *Brit. J. Med.* 5:2 (1948).
15. Nisioka, H. "Mutagenic Activities of Metal Compounds in Bacteria," *Mutat. Res.* 31:185 (1975).
16. Hood, R. D. and S. L. Bishop. "Teratogenic Effects of Sodium Arsenate in Mice," *Arch. Environ. Health* 24:62 (1972).
17. Fishbein, L. "Environmental Metallic Carcinogens: An Overview of Exposure Levels," *J. Toxicol. Env. Health* 2:77 (1976).
18. Langard, S. and T. Norseth. "A Cohort Study of Bronchial Carcinomas in Workers Producing Chromate Pigments," *Brit. J. Ind. Med.* 32:62 (1975).
19. Bigaliev, A. B., M. S. Elemesova and R. K. Bigalieva. "Chromosome Aberration Induced by Chromium Compounds in Somatic Cells of Mammals," *Tsitol. Genet.* 10:222 (1976).
20. Venitt, S. and L. S. Levy. "Mutagenicity of Chromates in Bacteria and Its Relevance to Chromate Carcinogenesis," *Nature* 250:493 (1974).
21. Correa, P., W. Haenszel, C. Cuello, S. Tannenbaum and M. Archer. "A Model for Gastric Cancer Epidemiology," *Lancet* 11:58 (1975).
22. Kier, L. D., E. Yamasaki and B. N. Ames. "Detection of Mutagen Activity in Cigarette Smoke Condensates," *Proc. Nat. Acad. Sci. USA* 71:4159 (1974).
23. Greenblatt, M., S. Mirvish and B. T. So. "Nitrosamine Studies: Induction of Lung Adenomas by Concurrent Administration of Sodium Nitrite and Secondary Amines in Swiss Mice," *J. Nat. Cancer Inst.* 46:1029 (1971).
24. Couch, D. B. and M. A. Friedman. "Interactive Mutagenicity of Sodium Nitrite, Dimethylamine, Methylurea and Ethylurea," *Mutation Res.* 31:109 (1975).

25. Eisenstark, A. and J. L. Rosner. "Chemically Induced Reversions in the Cyst Region of *Salmonella Typhimurium*," *Genetics* 49:343 (1964).

26. Frost, D. V. "The Two Faces of Selenium—Can Selenophobia be Cured," *CRC Crit. Rev. Toxicol.* 467 (1972).

27. Weisburger, J. H. and E. K. Weisburger. "Biochemical Formation and Pharmacological, Toxicological, and Pathological Properties of Hydroxylamines and Hydroxamic Acids," *Pharmacol. Rev.* 25:1 (1973).

28. Mukoi, F., L. Hawryluk and R. Shapiro. "The Mutagenic Specificity of Sodium Bisulfite," *Biochem. Biophys. Res. Commun.* 39:983 (1970).

AN INTRODUCTION TO ION CHROMATOGRAPHY

H. Small

Central Research—Physical Research Laboratory
Dow Chemical Co.
Midland, Michigan

ABSTRACT

A brief account of the conception, development, scope and limitations of the technique.

INTRODUCTION

Ion chromatography (IC) is a new technique that has been developed in recent years at The Dow Chemical Company, and it is finding increasing application in a number of different areas of analytical chemistry.[1-6] It employs some well-established ion exchange principles, but in a novel way, so that it is now possible to make use of electrical conductance as a means of quantifying the eluted ionic species. Since the bulk of this book is devoted to the applications of ion chromatography, I thought it fitting to present the principles on which the method is based and to give a very brief account of how and why the technique was developed.

Ion chromatography had its origins several years ago in a project then referred to as inorganic chromatography. It seemed at that time that there were many problems in chemistry whose solutions would be greatly expedited if one had a rapid method of analysis for common inorganic ions, such as the alkali metal and alkaline earth cations, and anions such as F^-, Cl^-, Br^-, I^-, SO_4^{2-}, SO_3^{2-}, NO_3^-, NO_2^- and PO_4^{3-}. It seemed that ion-exchange chromatography, because of its great separating power ought to yield a solution, and so eventually it did, but not until we had overcome a serious limitation in the ion-exchange chromatographic method. What is this limitation and how does it arise?

The development of any chromatographic method can be thought of as involving the solution of two problems—the separation problem and the analytical problem. The former involves selecting suitable stationary and mobile phases and elution conditions, etc., to effect good resolution of the sample species of interest, either from each other, or from any materials in the sample that are not of analytical interest but would interfere in the subsequent analysis. The analytical problem, on the other hand, involves quantitative measurement of the separated species that are eluted from the column. Some time ago, it might have been acceptable to think of these as two distinct and separate problems, but in recent years liquid chromatography has been developing toward high-speed separations coupled with continuous effluent monitoring by detectors, which can give essentially instantaneous analysis of the eluted species. This means, in turn, that separation and analysis are now intimately connected, and a separation scheme can be considered useful only to the extent to which it can be successfully coupled to a suitable monitoring device. It is in light of this present day practice that certain limitations of ion-exchange chromatography appeared.

It is intrinsic to ion-exchange chromatography that ionic solutions be used as eluents and, consequently, a successful detector must be capable of measuring the eluted species in this electrolyte background, often with a high degree of sensitivity. If the sample ions absorb in the visible or ultraviolet regions of the spectrum, then spectrophotometric methods may be applied to the analysis of the eluting ions. If the ionic species lack useful chromophores they can in some cases, by suitable reactions, be made to generate chromophores that can be measured spectrophotometrically. The ion-exchange analysis of amino acids is a classic and much-practiced example of this approach.

Beyond these examples, there are many ionic species that lack either chromophores or a generally applicable means of generating chromophores —most notably, those inorganic ions listed above. Therefore, to successfully determine these ions by chromatography, we felt the major problem to be that of developing a new means of detection.

From time to time, electrical conductance measurement has been proposed or tried as a means of analyzing ionic species in chromatographic eluents. Such a method is attractive for a number of reasons:

1. Since electrical conductance is a property common to all ionic species in solution, a conductivity monitor has the potential of being a universal monitor for all ionic species. This is a most desirable attribute in any chromatographic detector.

2. Conductance is simply related to species concentration for many ionic solutions.
3. Conductivity-measuring equipment is relatively simple and, as a result, will tend to be less subject to failure and should be relatively inexpensive.

Despite the obvious positive attributes of electrical conductance, its application as a detection principle has met with meager success. The principal reason for this lack of success is the problem presented by the eluent background. We can elaborate on this point by means of a simple illustrative example of determining sodium and potassium in a sample in which these are the only cations present. Excellent separation of these two ions may be obtained using a strong acid resin and an aqueous solution of hydrochloric acid as the eluent. However, attempts to use electrical conductance of the effluent as a means of monitoring for the sample ions will be frustrated by the presence of the HCl in the eluent—the high conductance of the eluent "swamps" the relatively low conductance due to the sodium and potassium peaks. Ion chromatography is the name given to the unique combination of means that was devised to cope with this problem of the eluent background.[1] A key part of IC is what we have chosen to call eluent suppression.

ELUENT SUPPRESSION

Again, an illustrative example serves to explain what is meant by eluent suppression. Using the earlier example of sodium and potassium analysis, if the effluent from the cation exchanger (the separating column) is passed into a strong base resin in the hydroxide form, two ion-exchange reactions will take place.

The HCl will be neutralized in the strong base resin, thus

$$HCl + Resin\ OH^- \rightarrow Resin\ Cl^- + H_2O$$

This is the suppression or stripping reaction and this second column is referred to as the suppressor or, as in the original publication,[1] the stripper column. The second reaction of note is the conversion of the NaCl and KCl peaks to equivalent peaks of NaOH and KOH.

As a result of the reactions in the suppressor column, the sample species are presented to a conductivity detector, not in a highly conducting background but rather in the very low conductivity of deionized water.

An analogous scheme can be envisaged for anion analysis, where sodium hydroxide is used as eluent in conjunction with a strong base resin in the separating column and a strong acid resin in the hydrogen form in the suppressor column. Deionized water is not always the product of eluent suppression. As will be seen later, other products are possible, but all have the property of low electrical conductivity, which is essential for the successful practice of IC.

EXHAUSTION AND REGENERATION OF
THE SUPPRESSOR COLUMN

Eluent suppression is a major key to the successful use of a conductance detector. A second very important consideration is the need to regenerate the suppressor bed, which has a limited capacity for modifying eluent. A too-frequent need to regenerate this bed would be an obvious drawback, so another important key to the successful practice of IC is the means that have been devised to make this regeneration step reasonably unobtrusive.

The problem of suppressor exhaustion can be expressed as a question; namely, how many samples can be analyzed before suppressor regeneration becomes necessary? The answer can, in turn, be approximately represented by the simple equation

$$N = \frac{V_B}{V_A} \cdot \frac{C_B}{C_A} \cdot \frac{1}{K_y^x}$$

where N = the number of samples injected during the suppressor's lifetime (assuming maximum utilization of the available lifetime); V_A and V_B are, respectively, the volumes of the separating and suppressor beds; C_A and C_B are, respectively, the specific capacities of the separating and suppressor resins; and K_y^x is a selectivity coefficient, which describes the affinity of the separating resin for the eluting ion Y relative to the ion X, which, in a series to be analyzed, has the greatest affinity for the separating resin. The higher the affinity of the resin for the displacing ion, Y, the smaller is the value of K_y^x. Since the objective is to maximize N, the equation suggests various means of doing this, and these will now be treated individually.

The Factor V_B / V_A

Although N may be increased by increasing the value of V_B/V_A, an upper limit of 10 is recommended for this quantity to avoid loss of resolution by mixing in the void volume of an excessively large suppressor column. Other problems can arise, especially for partially ionized electrolytes, when a suppressor bed is used that is too large relative to the separating bed.[1] Normally, we use separating and suppressor columns that are equal in volume.

The Factor C_B/C_A

While the useful variation of V_B/V_A is quite limited, there is much more latitude permitted in varying C_B/C_A. The specific capacity of the suppressor bed is made as large as possible by using commercially available resins of moderate-to-high crosslinking. The specific capacity of the separating bed is kept small by using resins with superficial functionality (pellicular resins). A lower limit on the capacity of this column is determined by the need to avoid overloading by the ions of the injected sample. In practice, we find that useful capacities for the separating resins are in the range of 0.005 to 0.002 meq/ml.

For cation analysis by IC, low-capacity resins are prepared by surface sulfonation of styrene divinylbenzene copolymer beads.[1] Low-capacity anion exchangers are prepared by a surface agglomeration method, where finely divided anion exchange resin is contacted with surface-sulfonated S/DVB copolymer. The small particles of anion exchanger are held tenaciously on the oppositely charged surface of the sulfonated beads. In the original work, the fine anion exchanger was prepared by grinding Dowex® 2 resin followed by sedimentation refining. More recently, Jitka Solc, of our laboratories, has synthesized very small particle size distribution by quaternizing emulsion copolymers of vinylbenzyl chloride and divinylbenzene. Figure 2.1 shows scanning electron micrographs of the latex form ion exchangers attached to the surface-sulfonated substrate particles. Columns may be prepared *in situ* by injecting a dilute suspension of the ion-exchanging latex into a column that is already filled with the surface-sulfonated resin. The anion-exchange capacity of this surface-agglomerated separating resin is controlled by the amount and concentration of suspension injected and by the size of the individual latex particles. The selectivity of the resin is controlled in the usual manner; that is, by varying the divinylbenzene/vinyl benzyl chloride ratio. The electrostatic attachment of the monolayer of latex yields resins of very stable capacity. Even quite concentrated electrolyte solutions will not displace a significant amount of the bound monolayer. Furthermore, the surface-agglomerated resins are stable over a wide range of pH, in which respect they are superior to glass or silica-based pellicular resins, which are quite unstable at the high pH levels normally employed in the analysis of anions by IC.

The Factor $1/K_y^x$

The lifetime of the suppressor is prolonged by using displacing ions (Y) of high affinity relative to the sample ions. This suggests a large choice in eluent ions, but, in fact, the choice is limited by three further requirements:

A

—— 50 μm ——

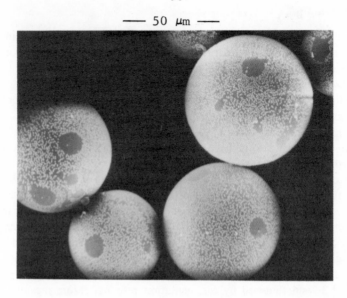

B

—— 5 μm ——

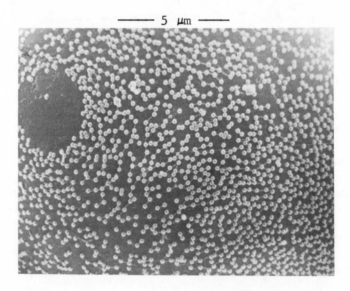

Figure 2.1 A. Scanning electron micrograph showing 3600-Å-diameter anion exchange particles adhering to the surface of the substrate beads.
B. A larger magnification of an area of one of the beads.

1. that the eluent conductance be amenable to suppression by some convenient reaction of the eluent ions with the resin in the second column;
2. that the sample ions not be subject to an adverse reaction (*e.g.*, removal) in the suppressor bed; and
3. that the affinity of the displacing ion, Y, not be so high that the sample ions elute too rapidly without adequate resolution.

Despite these limitations, a number of eluents of both anions and cations have been successfully used in IC. They are listed in Table 2.1 along with the appropriate suppressor reaction.

Table 2.1

Eluent	Eluent Ion	Suppressor Resin	Products of Suppressor Reaction
Anion Analysis			
NaOH	OH^-	R^-H^+	$R^-Na^+ + H_2O$
Na phenate	ϕO^-	R^-H^+	$R^-Na^+ + \phi OH$
$Na_2CO_3/NaHCO_3$	CO_3^{2-}/HCO_3^-	R^-H^+	$R^-Na^+ + H_2CO_3$
diNa-glutamate	glut.$^{2-}$	R^-H^+	$R^-Na^+ + R^-$ glut. H^+
Cation Analysis			
HCl	H^+	R^+OH^-	$R^+Cl^- + H_2O$
$AgNO_3$	Ag^+	R^+Cl^-	$R^+NO_3^- + AgCl\downarrow$
$Cu(NO_3)_2$	Cu^{2+}	$R\ NH_2$	$R\ NH_2 \cdot Cu(NO_3)_2$
Pyridine $\cdot$ HCl	PyH^+	R^+OH^-	$R^+Cl^- + Py + H_2O$
Aniline $\cdot$ HCl	ϕNH_3^+	R^+OH^-	$R^+Cl^- + \phi NH_2 + H_2O$
p-PDA* $\cdot$ 2 HCl	p-PDAH$_2^{2+}$	R^+OH^-	$R^+Cl^- + p$-PDA $+ H_2O$

For anion analysis, the hydroxide ion is ideal from the viewpoint of suppression, but useful only for elution of less tightly bound monovalent anions such as, for example, F^-, Cl^- and acetate. For more tightly bound sample ions, especially polyvalent ions such as SO_3^{2-}, SO_4^{2-} and PO_4^{3-}, more potent displacing ions must be used. The phenate ion is very useful, and several examples have been described which illustrate its application.[1] More recently, we have found various combinations of HCO_3^-, CO_3^{2-} and OH^- to be very effective as eluents.[2] Such systems exploit the superior affinity of the divalent CO_3^{2-}, and the fact that the dilute carbonic acid formed in the strong acid suppressor bed is of relatively low conductivity. The higher displacement potential of CO_3^{2-} is not the only advantage of these systems, since the use of various combinations of HCO_3^-, CO_3^{2-} and OH^- affords a means to vary pH in the separating column over a greater range than is possible with either the hydroxide or phenate eluent systems. This proves advantageous in the elution of species such as orthophosphate, whose charge (HPO_4^{2-}, PO_4^{3-}) and, hence, ease of displacement, is pH-sensitive.

For cation analysis, the hydronium ion, H^+, has already been discussed and is an ideal elution ion as it forms water in the suppressor bed. On the other hand, its usefulness is limited by its relatively low affinity for the separating resin, and such high concentrations of H^+ are needed to displace more tightly bound cations, that suppressor lifetime is in many cases reduced to impractically low values. Eluent cations of greater potency than H^+ therefore must be used. Inorganic cations such as Ag^+ and Cu^{2+} have been successfully used for the separation of more tightly bound sample ions such as the alkaline earth ions. Suppression of the Ag^+ is achieved by precipitating Ag^+ as silver chloride on an anion exchanger in the chloride form while the copper salt is removed from the eluent by complexing with a polyamine resin. In both cases, suppression is achieved by stripping the electrolyte completely from the eluent. Mineral acid salts of certain organic amines such as pyridine, aniline, and p-phenylene diamine are very effective eluents due to the high affinity of the protonated amines for the sulfonated styrene-based resin. In these cases, suppression of the eluent conductivity is brought about not by removal of the electrolyte, but by its conversion to a much less conducting form in the suppressor column. A strong base resin converts the amine hydrochlorides to their respective free bases which, being very weak, are feebly dissociated and contribute very little to the effluent conductance.

SCOPE AND LIMITATIONS OF IC

The principal appeal of IC is its ability in a single elution to rapidly analyze a mixture whose components have quite diverse chemical properties. The separation of Figure 2.2 serves to illustrate this point— -F^-, Cl^-, PO_4^{3-}, NO_3^- and SO_4^{2-} may all be analyzed by this method in about 20 minutes, whereas analysis of the individual ions by wet chemical methods would be more laborious. Since it is a chromatographic method, IC has the ability to analyze a low concentration of one ion in a preponderance of another and is able, therefore, to analyze ions of interest against a background of ions that may cause interference in other methods of analysis.

High sensitivity and rapid analysis are two positive attributes of IC. Sensitivity of detection is determined by the sharpness of the eluted peaks and their conductance level above the background conductance. Thus, sensitivity will be greater in a deionized water effluent than in a dilute

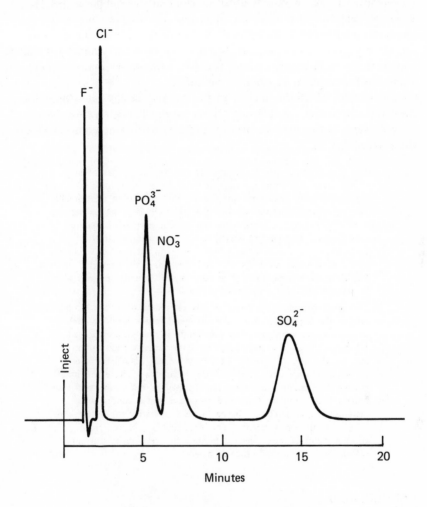

Figure 2.2 A typical anion chromatogram.

carbonic acid or phenol effluent. On the other hand, the superior eluting power of carbonate or phenate systems as compared to sodium hydroxide may, in many cases, more than counterbalance the advantage of the lower background conductance of the hydroxide system.

IC is routinely capable of analyzing in the part per million level for a great many ions. Analysis at the 1 part in 10^8 level has been achieved by Stevens[5] in the cases of Na^+, K^+, Cl^- and SO_4^{2-}.

The speed of IC analysis is usually related to the complexity of the analytical sample. In general, the more ions the sample contains the longer the elution, even though not all contained ions may be of analytical interest. One minute per ion is not uncommon for samples containing two to three easily separated ions, and five minutes per ion is easily attainable for many more complex mixtures.

In principle, IC can be applied to the analysis of any ionic species since the detector, a conductivity cell, is, within limits, sensitive to all ionic species. There are two main considerations that determine limits to the applicability of IC:

1. The first is the ability to adequately resolve the species of interest in the separating column. Certain ionic species may have such similar ion-exchange selectivity coefficients that resolution is attainable only with prolonged elution through relatively large separating columns. This may put such a burden on the suppressor that few samples may be analyzed during the lifetime of the suppressor. Considerations such as daily sample load and relative convenience and cost of other methods will then determine the preferred method. Ion chromatographic analysis of very tightly bound ions may also be limited by short suppressor lifetimes.

2. The other consideration concerns adverse reactions of the sample ions in the suppressor bed. Certain metal ions cannot be analyzed with IC systems because they precipitate in the suppressor bed. Likewise, IC has low sensitivity for determining certain anions because the ions of interest form very weakly dissociated species in the suppressor bed. The inability to analyze silicate ion by IC is such an example—the silicic acid formed in the strong acid suppressor bed is too weak an acid to give a useful conductance response in the detector. IC has a low sensitivity for determining CO_3^{2-} for the same reason. This drawback is a virtue from another point of view, since it permits the very successful application of carbonate solutions as eluents.

CONCLUSIONS

Ion chromatography is a practical ion exchange method for analyzing both cations and anions. Its use of conductimetric detection gives it universal applicability to all ionic species and special utility for ions for which no other convenient detection method is available. The method is particularly attractive for anions, in light of the varied and often laborious techniques required for their analyses.

REFERENCES

1. Small, H., T. S. Stevens and W. C. Bauman. "Novel Ion Exchange Chromatographic Method Using Conductimetric Detection," *Anal. Chem.* 47:1801 (1975).
2. Small, H. and J. Solc. "Ion Chromatography—Principles and Applications," Proceedings of An International Conference on "The Theory and Practice of Ion Exchange," Cambridge, England, 1976.
3. Anderson, C. "Ion Chromatography: A New Technique for Clinical Chemistry," *Clin. Chem.* 22:1424 (1976).
4. Mulik, J., R. Puckett, D. Williams and E. Sawicki. "Analysis of Nitrate and Sulfate in Ambient Aerosols," *Anal. Letters* 9(7):653 (1976).
5. Bouyoucos, S. A. "The Determination of Ammonia and Methylamines in Aqueous Solutions by Ion Chromatography," *Anal. Chem.* 49:401 (1977).
6. Colaruotolo, J. F. and R. S. Eddy. "Determination of Chlorine, Bromine, Phosphorus and Sulfur in Organic Molecules by Ion Chromatography," *Anal. Chem.* 49:884 (1977).
7. Stevens, T. S. "Ion Chromatography Applications," 2nd Annual Meeting of Federation of Analytical Chemistry and Spectroscopy Societies, Indianapolis, Indiana, 1975.

3

ION CHROMATOGRAPHIC DETERMINATION
OF ATMOSPHERIC SULFUR DIOXIDE

J. D. Mulik, G. Todd, E. Estes, R. Puckett, E. Sawicki

Environmental Sciences Research Laboratory
U.S. Environmental Protection Agency
Research Triangle Park, North Carolina 27711

D. Williams

Northrop Services, Inc.
Environmental Sciences Group
Research Triangle Park, North Carolina 27709

ABSTRACT

A new method has been developed for the 24-hour quantitative collection and analysis of atmospheric SO_2. This is based on a method in which atmospheric SO_2 is bubbled through a reagent that converts SO_2 to sulfate, followed by ion chromatographic analysis.

Preliminary studies indicate that the reagent has a collection efficiency of approximately 100% over the range of 24 to 1291 $\mu g/m^3$ SO_2 with no apparent interferences and that the SO_2-derived sulfate is easily assayed by ion chromatography. The collecting reagent is stable for 30 days both before and after collection of SO_2 and at temperatures as high as $100°F$. The inadequacies of the Federal Register Reference Method for SO_2 are also described.

INTRODUCTION

The primary source of atmospheric SO_2 is from the combustion of coal and petroleum. Since sulfur dioxide is one of the six major pollutants for which a national ambient air quality standard has been set by the federal government, it is imperative that a reliable and accurate analytical method be available for SO_2. Of the many methods available for the collection and analysis of SO_2, the most common procedure is the Federal Reference Method,[1] which is a modified West Gaeke method.[2]

The Federal Reference Method involves the collection of ambient SO_2 by bubbling air through an aqueous solution of 0.04 M potassium tetrachloromercurate (TCM) for 24 hours to form a sulfito complex. The sulfito complex is then reacted with formaldehyde and pararosanaline and assayed colorimetrically. Scaringelli et al.,[3,4] in their attempts to improve the West Gaeke method, reported that the TCM solution had a 1 to 3% per day decay at 25°C depending on the concentration. More recently, Fuerst et al.[5] quantified the decay rate of the Federal Reference Method for SO_2 between 20° and 50°C. The rate of decay increases fivefold for every 10°C rise in temperature, as shown in Table 3.1.

Table 3.1 West-Gaeke SO_2 Reference Method: Effect of Temperature on Percent Decay Per Day[a]

Temperature, °C	Percent Loss Per Day
20	0.9
30	5.0
40	25.0
50	73.6

[a]EMSL Report 600/4-76-024, R. G. Fuerst et al. (May 1976).

Since it has been common practice to collect the SO_2 in sampling boxes thermostatted at 35°C and since the samples are sent back to the laboratory in nontemperature-controlled containers, the SO_2 measurements that have been made under these conditions are suspect.

Although Fuerst et al.[5] showed that if the temperature was maintained at 12° ± 5°C during and after collection of SO_2 that reliable SO_2 data could be obtained with the Federal Reference Method, we believed that a new method was still needed. The primary reason was that it would be difficult and quite expensive to maintain the low temperature during and after collection. Even if the temperature could be controlled, a difficult colorimetric procedure, a highly toxic chemical would still be used for analysis. From our discussions with personnel involved with SO_2 collection and analysis, it was understood that they would prefer to keep the bubbler network for SO_2 without any major changes, such as switching to expensive thermoelectric coolers. We suggested that SO_2 could be collected in a H_2O_2[6] solution that would convert it to sulfate, after which it could be easily analyzed with a relatively new technique called ion chromatography (IC).

Ion chromatography was first described by Small et al.[7] for the analysis of various anions and cations, and consists of ion-exchange chromatography, eluent suppression and conductimetric detection. Small's novel technique of eluent suppression allowed the use of the simple and universal conductivity detector. Eluent suppression is carried out by a second ion-exchange column that reduces or suppresses the unwanted eluent ions without affecting the eluting ion species.

Mulik et al.[8] recently showed that ion chromatography is a highly sensitive and selective method for the analysis of nitrate and, particularly, sulfate in ambient aerosols. It followed that if we could find a stable and efficient collecting solution to convert the SO_2 to sulfate we would have a simple, reliable-sensitive and selective measurement for SO_2 with the ion chromatograph.

The purpose of this chapter is to present data on the successful application of a dilute solution of hydrogen peroxide as an absorber for the manual 24-hour collection of SO_2 using ion chromatography as the analytical technique.

EXPERIMENTAL

In evaluating any method for the analysis of an air pollutant, accurate, static and dynamic concentrations of the pure pollutant must be available.

Dynamic

National Bureau of Standards-certified SO_2 permeation tubes were used as dynamic calibration standards. SO_2 permeates from this tube at a constant temperature-dependent rate. Although NBS had certified the tube, the rate of SO_2 effusion was verified gravimetrically in our laboratory. Equation 1 was used to calculate the theoretical concentration of SO_2 in the generated test atmospheres.

$$C_T = \frac{PR \times 1000}{TDA} \tag{1}$$

C_T = theoretical concentration of SO_2 $\mu g/m^3$
PR = permeation rate of SO_2 $\mu g/min$
TDA = total dilution air liters/min
1000 = conversion of liters to m^3

Static

The ion chromatograph was calibrated with standard sulfate solutions prepared from certified 0.1 N sulfuric acid (Fisher*).

Detailed calibration curves were run periodically. However, for the data presented herein the single-point external standard method was used. The sample is run first to estimate the species concentration. Then a closely matching standard is assayed. Quantitation is obtained by the ratio of the peak area of the standard to the peak area of the sample as will be discussed.

Absorbing Solution

Hydrogen peroxide -

1 ml of 30% H_2O_2 + 49 ml of deionized H_2O + 5 μl of 0.6 N HCl

CALIBRATION PROCEDURE

SO_2 Test Atmosphere Generator

Figure 3.1 is a schematic of the apparatus used for the generation of known SO_2 test atmospheres. The apparatus consisted of a NBS SO_2 permeation tube housed in a condenser, which was supplied with water at 25 ± 0.1°C from a constant temperature bath; clean dry N_2 continuously passed over the SO_2 device at 50 ml/min⁻¹. Additional purified air was added at known flow rates to obtain the desired concentrations of SO_2. Ideally, the SO_2 test atmosphere should be monitored continuously by a SO_2 flame photometric monitor to serve as an indicator of possible upsets in flow, temperature, etc., during the sampling period.

Sampling Procedure

The basic sampling unit shown in Figure 3.2 was used to collect the air samples. The unit consisted of a restricted orifice bubbler immersed in the peroxide-absorbing solution, a trap to remove moisture, a critical orifice (#27 hypodermic needle), and a vacuum pump capable of pulling 16 in. of Hg. The pump pulls air through the abosrbing solution at approximately 200 ml/min, as controlled by the critical orifice. Five of

---

*Any mention of commercial products does not constitute endorsement by the United States Environmental Protection Agency.

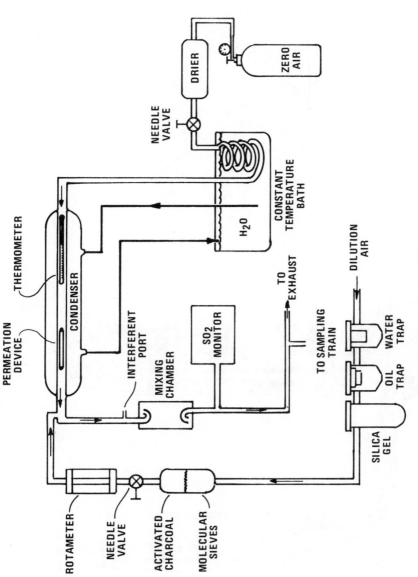

Figure 3.1 SO$_2$ test atmosphere generator.

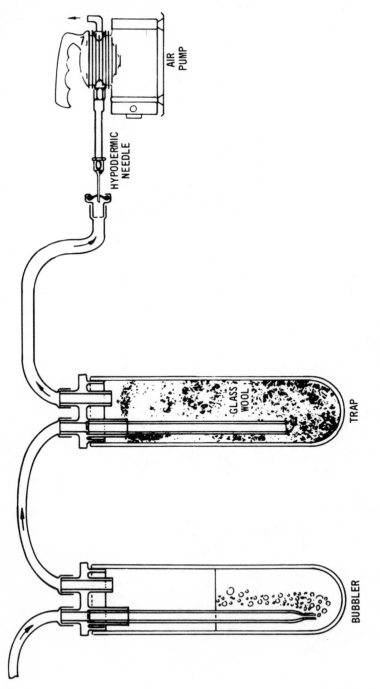

Figure 3.2 Single bubbler unit.

these sampling units were assembled in parallel by means of a glass manifold, to determine the precision of the method. Fifty milliliters of the peroxide-absorbing reagent were placed in each of the five bubblers in the sampling train (Figure 3.3). The flow rate through each of the bubblers was measured with a soap bubble flow meter, and the total dilution flow for the permeation device was measured with a wet-test meter. The sampling train was then attached to the SO_2 test generator. At the end of a sampling period of 20 to 24 hours, the flow rates were remeasured.

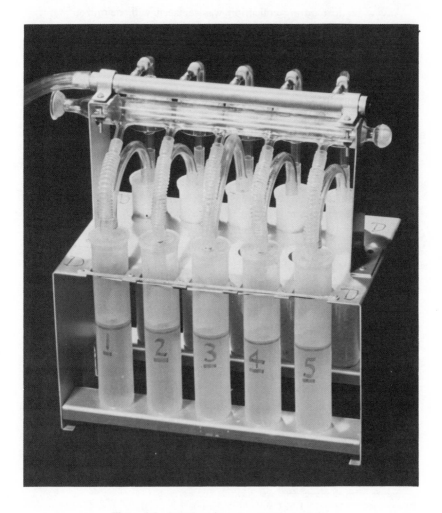

Figure 3.3 Bubbler train—5 units in parallel.

The volume of air sampled in m^3 was calculated from Equation 2.

$$V = \frac{(F_1 + F_2)}{2} \times T\ 10^{-6} \tag{2}$$

V = volume of air, m^3

F_1 = measured flow before sampling, ml/min

F_2 = measured flow rate after sampling, ml/min

T = sampling, time, min

10^{-6} = conversion of ml to m^3

Any water lost during collection was replaced with deionized water before ion chromatographic analysis was performed on aliquots from each tube.

Analytical

Analysis of the SO_2 as sulfate in the peroxide solution was done with an ion chromatograph. A schematic of the Model 10 ion chromatograph (Dionex Corp., Palo Alto, California) is shown in Figure 3.4. It is basically a low-pressure liquid chromatograph consisting of a separator or analytical column, suppressor column, solvent reservoirs, inject valve, 0.5-ml sample loop, two Milton Roy pumps, conductivity detector and a valving system to direct the flow through various parts of the instrument. All valves are air-actuated Teflon®* slider valves.

For the analysis of SO_2 as sulfate, the separator column contains a strong basic anion exchange resin, and the suppressor column contains a strong acid resin in the hydrogen form.

The eluent was 0.003 M $NaHCO_3$ and 0.0024 M Na_2CO_3 at a flow rate of 3.8 ml/min^{-1} and pump pressure of 500 PSIG.

Two-milliliter aliquots of the peroxide collection solution are injected by syringe into the ion chromatograph. The sample loop on the inject valve contains 0.5 ml. The 2-ml volume is necessary to insure proper flushing of the inject valve loop and lines.

A typical ion chromatogram is shown in Figure 3.5 of a H_2O_2-absorbing solution used to collect SO_2 that was generated at 24 $\mu g/m^3$. Only 0.25 m^3 was collected, which means that the 50 ml of absorber had approximately 0.12 μg SO_2/ml. Since the sample loop is 0.5 ml, the peak on the ion chromatogram in Figure 3.5 represents 0.06 μg of SO_2, which would indicate that the H_2O_2-IC method has adequate sensitivity to measure SO_2 in ambient air samples.

---

*Registered trademark of E. I. duPont de Nemours & Company, Inc., Wilmington, Delaware.

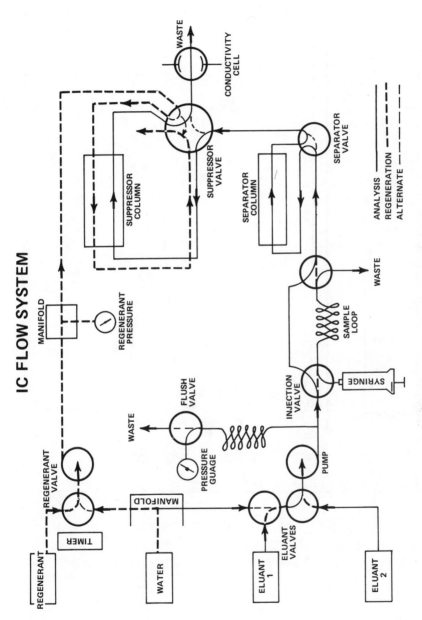

Figure 3.4   Schematic of model 10 ion chromatograph.

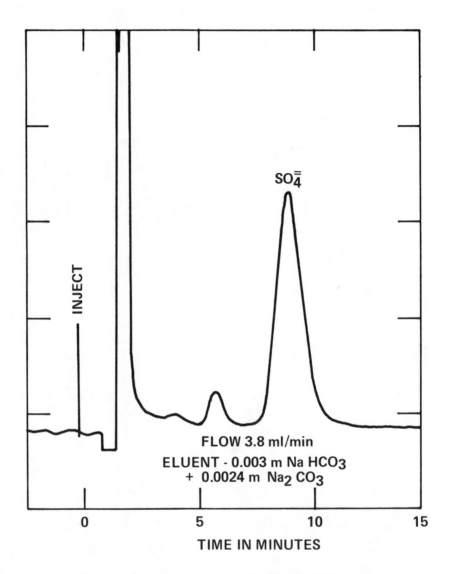

Figure 3.5 Typical ion chromatogram of H_2O_2 bubbler solution.

The quantity of sulfate, in μg/ml, found in the peroxide-absorbing solution was calculated from Equation 3.

$$A = \frac{C_{std}}{PA_{std}} \times PA_{unk} \tag{3}$$

A = concentration of unknown, $\mu g\ SO_4^=/ml$

C_{std} = concentration of standard, $\mu g\ SO_4^=/ml$

PA_{std} = peak area of standard

PA_{unk} = peak area of unknown

Peak areas were determined by the peak height times the peak width at half peak height. It would be much easier and more accurate if an electronic integrator could be used, but difficulties were encountered in interfacing the electronic integrator with the ion chromatograph. The actual quantity of SO_2 in $\mu g/m^3$ was calculated from Equation 4.

$$C_F = \frac{0.67 \times 50A}{V} \tag{4}$$

C_F = concentration of SO_2 found, in $\mu g/m^3$

0.67 = conversion factor for converting $SO_4^=$ to SO_2

50 = volume of collecting solution, ml

V = volume of air sampled, m^3

A = concentration, $\mu g\ SO_4^=/ml$ (obtained from Equation 3)

RESULTS AND DISCUSSION

All percentage recovery data reported herein depend solely on the amount of SO_2 recovered as sulfate and do not involve any stoichiometric factors. Table 3.2 shows the average collection efficiency of the peroxide absorber over the range of 24 to 1291 $\mu g\ SO_2/m^3$ to be essentially 100%. Figure 3.6 contains the data from Table 3.2 plotted in $\mu g\ SO_2/m^3$ found vs $\mu g\ SO_2/m^3$ generated. Each of the points is the average of five determinations. The correlation coefficient of 0.989 could and would have been better except for problems with the constant temperature bath for the permeation tube during our 170 to 200 $\mu g\ SO_2/m^3$ runs. In spite of this, the correlation was quite good.

Stability and Interference Studies

Essentially 100% recovery, as shown in Table 3.3, was obtained with the peroxide-absorbing solution even 30 days after the solution was prepared.

The IC analysis of the SO_2 immediately after collection and analysis of the same solution 30 days later are shown in Table 3.4.

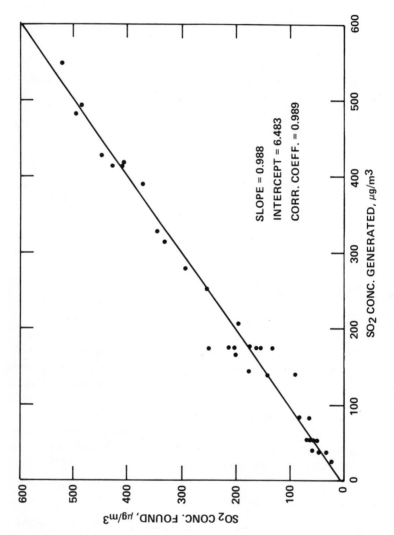

Figure 3.6 Plot of SO₂ found by IC vs SO₂ generated.

Table 3.2   Percent Recovery of SO_2 - H_2O_2 Collection + IC Analysis

SO_2 Generated ($\mu g/m^3$)	SO_2 Found ($\mu g/m^3$)	Percent Recovery	$S_{(REL)}$ (%)
24.3	23.0	94.0	23.0
37.0	36.6	99.0	9.0
54.0	59.0	109.0	10.1
84.0	83.7	100.0	2.0
139.0	141.6	102.0	5.3
175.0	203.7	116.0	9.7
253.0	256.0	101.0	2.3
278.0	293.0	105.0	5.5
416.0	408.0	98.0	2.4
548.0	520.0	95.0	1.1
804.0	775.0	96.0	2.1
1067.0	1133.0	106.0	0.8
1291.0	1357.0	105.0	1.7
		$\overline{X}$ = 102.0	$\overline{X}$ = 5.7

$^a S_{(REL)}$ % = $\dfrac{\text{Standard Deviation}}{\text{Average Recovery}}$ x 100.

Table 3.3   Stability of H_2O_2 Before SO_2 Collection

Age of Solution, Days	Percent Recovery by IC 139 μg SO_2/m^3
1	98
30	102

Table 3.4   Stability of H_2O_2 Solution After SO_2 Collection

	Percent Recovery by IC	
	178 μg SO_2/m^3	804 μg SO_2/m^3
Initial Assay	98	96
30 Days Later	98	98

These data indicate that the peroxide solution has excellent stability both before and after collection of the SO_2. No attempt was made to protect the solutions from light or to keep them cold.

The data in Table 3.5 show that temperature, from approximately 5 to $40°C$ had no effect on the collection efficiency of the peroxide-absorbing solution. Ozone, hydrogen sulfide, methyl mercaptan, carbonyl sulfide, particulate sulfate and relative humidity were studied as possible interferences. The sulfur-containing species were thought to be the most likely positive interferents.

Table 3.5   Effect of Temperature on H_2O_2 Collection and IC Analysis

Temperature of H_2O_2	μg SO_2/m^3 Generated	μg SO_2/m^3 Found by IC	Percent Recovery
$5°C$[a]	86	89	104
$40°C$[a]	397	379	95
$40°C$[b]	175	163	94

[a] During 24-hour collection.
[b] Heated for 30 days before collection.

Moisture, when using a 5 μ prefilter to remove particulate sulfate, could be a negative interference due to the SO_2 being collected on the wet filter. However, the 5 μ Teflon prefilter will pass SO_2 quantitatively at low concentrations and high relative humidity (100%), as shown in Table 3.6.

Table 3.6   Effect of Prefilter and Moisture on Collection Efficiency

	Percent Recovery	
	174 μg SO_2/m^3	50 μg SO_2/m^3
Prefilter (5 μg Teflon Millipore)	95	95
Prefilter + 100% Relative Humidity	92	95

As can be seen in Table 3.7, no interferences from O_3, H_2S, $CH_3 SH$ and COS were found at the levels studied. These interferent levels are much higher than concentrations found for most 24-hour ambient air samples. NO_2 will not be converted to nitrate with this particular absorbing solution. Since it would be advantageous to do gaseous NO_2 as nitrate with IC, further research is being carried out on other absorbing reagents that can convert NO_2 to nitrate.

It has been postulated that SO_2 can also be lost due to conversion of the SO_2 to sulfate by the particulate that builds up on the prefilter during the 24-hour sampling period. Since the prefilter will be changed

Table 3.7   Interference Studies

Concentration of Interferent ($\mu g/m^3$)		Concentration of SO_2 $\mu g/m^3$	Percent Recovery
O_3	(980)	139	102
O_3	(980)	54	104
H_2S	(120)	132	97
CH_3SH	(200)	174	95
COS	(71)	208	100

daily and since the flow rate through the prefilter is only 200 cc/min^{-1}, the SO_2 loss on the prefilter is expected to be minimal.

Data in Tables 3.8 and 3.9 show a comparison of the H_2O_2 - IC technique with the Federal Reference (TCM) - colorimetric technique on standard SO_2 samples generated in the laboratory.

Table 3.8   Comparison of Ion Chromatographic Method
With the West Gaeke/Colorimetric Method

	μg SO_2/m^3		Percent Recovery	
Generated	Found by IC	Found by TCM	IC	TCM
25	23.4	25.8	94	103
25	26.0	32.0	104	128
25	26.0	28.8	104	115
25	27.0	29.9	108	115
			$\bar{X}$ = 103	X = 115

Table 3.9   Comparison of Ion Chromatographic Method
With the Federal Reference (TCM) Colorimetric Method

	μg SO_2/m^3		Percent Recovery	
Generated	Found by IC	Found by TCM	IC	TCM
400	416	397	104	99
402	403	405	100	100
402	402	390	100	97
402	410	384	102	95
			$\bar{X}$ = 101	$\bar{X}$ = 98

The comparison data on the standard samples obtained from the SO_2 generator (Tables 3.8 and 3.9) appear to indicate that the TCM colorimetric method performs almost as well as the H_2O_2-IC method in the laboratory, except at extremely low concentrations.

A comparison of the two techniques of a 24-hour ambient air sample taken in the EPA annex parking lot is shown in Table 3.10. Three of the bubbler tubes contained peroxide, and two bubblers contained TCM. The ambient temperature during the sampling period reached a high of $95°F$ during the day and a low of $65°F$ through the night. This may explain why the TCM colorimetric method showed zero SO_2. The H_2O_2-IC method detected an average of 29 $\mu g/m^3$ (11 ppb), which is a higher SO_2 value than is generally found in this area. Since we had no prefilter on the bubbler train, it is possible that the high number was due to particulate sulfate.

Table 3.10   Ambient Air EPA Annex Parking Lot

SO_2 $\mu g/m^3$	Method
1. 31.4	H_2O_2-IC-No Prefilter
2. 31.7	
3. 26.0	
4. 0	TCM-Colorimetric
5. 0	

Table 3.11 shows similar data that were collected on the EPA annex roof. However, during this sample collection a 5 μ Teflon millipore prefilter was used. Again the TCM colorimetric method was unable to detect SO_2 and the H_2O_2-IC method found about 10 μg SO_2/m^3 (4 ppb) which is typical of SO_2 concentrations found in this area.

Table 3.11   Ambient Air EPA Annex (Roof)

SO_2 $\mu g/m^3$	Method
1. 9.5	H_2O_2-IC $+$ Prefilter
2. 8.5	
3. 11.5	
4. 0	TCM-Colorimetric
5. 0	

CONCLUSIONS

The Federal Reference Method for the collection and analysis of atmospheric SO_2 has been shown to be inadequate primarily because of the temperature decay of the sulfito mercury complex, which infers that most data collected by the Federal Reference Method is suspect. The hydrogen peroxide collection and ion chromatographic analysis for SO_2 in ambient air presents several attractive features: essentially 100% collection efficiency; no apparent interferences; stability before, during and after collection of SO_2 at low and high temperatures; and use of a solution that is nontoxic. These features make the H_2O_2 procedure for the collection and IC analysis of SO_2 an excellent candidate for the replacement of the Federal Register Reference Method or, at the very least, an equivalent SO_2 method.

From the data presented herein, it appears that the H_2O_2-IC method is superior to the Federal Reference Method. However, before it could be made a Federal Reference Method or an equivalent method, much more comparison data must be obtained.

REFERENCES

1. Title 40, CFR Part 50, Appendix A. "Reference Method for the Determination of Sulfur Dioxide in the Atmosphere," 36 *FR* 22, 385 (November 25, 1971).
2. West, P. W. and G. C. Gaeke. "Fixation of Sulfur Dioxide as Sulfito Murcurate III and Subsequent Colorimetric Determination," *Anal. Chem.* 28:1816 (1956).
3. Scaringelli, F. P., B. E. Saltzman and S. A. Frey. "Spectrophotometric Determination of Atmospheric Sulfur Dioxide," *Anal. Chem.* **39**:1709 (1967).
4. Scaringelli, F. P., L. Elfers, D. Norris and S. Hochheiser. "Enhanced Stability of Sulfur Dioxide in Solution," *Anal. Chem.* 42:1818 (1970).
5. Fuerst, R. G., F. P. Scaringelli and J. H. Margeson. "Effect of Temperature on Stability of Sulfur Dioxide Samples Collected by the Federal Reference Method," Environmental Monitoring Series EPA-600/4-76-024 (May 1976).
6. Hochheiser, S. "Determination of Sulfur Dioxide: Hydrogen Peroxide Method," Environmental Health Series Air Pollution. "Selected Methods for the Measurement of Air Pollutants," USDHEW Public Health Service Publication No. 999-AP-11, Cincinnati, Ohio; May, 1965.
7. Small, H., T. S. Stevens and W. C. Bauman. "Novel Ion Exchange Chromatographic Method Using Conductimetric Detection," *Anal. Chem.* 47:1801 (1975).
8. Mulik, J. D., R. Puckett, D. Williams and E. Sawicki. "Ion Chromatographic Analysis of Sulfate and Nitrate in Ambient Aerosols," *Anal. Letters* 9(7):653 (1976).

DISCUSSION

Have you tried any other levels of H_2O_2 concentrations for collection?

We used as much as 4 ml of 30% H_2O_2 and we didn't see any difference, so we stayed with the 1 ml of 30% H_2O_2.

What kind of temperature problems were you having and what temperature were you controlling?

The temperature of the bath circulator that controlled the permeation tube went up. It had nothing to do with the ion chromatograph.

Did you use an SO_2 monitor during collection?

No. I wanted to use it, but we had none available during the study. I pointed out that it would have been good to use one to monitor what we were generating over the 24-hour period to see if there were any upsets in flows or temperature.

Couldn't SO_2 monitors be used for analysis in the field?

There are too many sampling sites for that to be practical, particularly at $4000 to $5000 per SO_2 monitor.

What can you say about the relative speed of the two different analyses when you are doing SO_2 as SO_4?

We shortened the analysis time by increasing the eluent flow since we are not interested in looking at the front end of the ion chromatogram. It takes about 6 minutes to do SO_2 as SO_4 by this technique. The TCM is generally done with an Automated Technicon, which can probably do a sample in less than five minutes. However, I would rather have fewer data that I could trust than a lot of data that may be suspect.

You mentioned that particulate sulfate can be removed with a prefilter with minimal loss of SO_2, because of the low flow rate used in your sampling system. Have you performed any experiments to verify this theory?

Yes, we added 200 μg of particulate to a prefilter and sampled SO_2 at 259 $\mu g/m^3$ through the filter. Our ion chromatographic analysis of the peroxide solution showed that we recovered 257 $\mu g/m^3$ SO_2, which is better than 99% recovery. This kind of recovery along with the fact that the prefilter will be replaced daily indicated that the use of a prefilter will not affect the recovery of SO_2.

4

ION CHROMATOGRAPHIC ANALYSIS OF
AMMONIUM ION IN AMBIENT AEROSOLS

J. D. Mulik, E. Estes and E. Sawicki
Environmental Sciences Research Laboratory
U. S. Environmental Protection Agency
Research Triangle Park, North Carolina 27711

ABSTRACT

This chapter describes the ion chromatographic analysis of ammonium ion (NH_4^+) in ambient aerosols collected on Hi-Vol and dichotomous filters. Analytical conditions and data on sensitivity, selectivity, accuracy and repeatability are discussed.

In future studies, this method will be applied to the analysis of pollutants of importance in carcinogenic research such as aliphatic and aromatic amines.

INTRODUCTION

Ammonia is a natural constituent of the atmosphere and generally exists at a concentration below the level at which it is hazardous to humans. Most of the ammonia in the atmosphere is produced by natural biological processes, largely from the decomposition of organic waste material. Man contributes a comparatively small portion of ammonia to the atmosphere, mainly through combustion and industrial processes involved in the production or use of ammonia.

Although an air quality standard for ammonia has not been set in the United States, there is still a need for a simple and accurate method for the analysis of ammonia gas and ammonium ion in airborne particulates. The interest in ammonium ion is primarily because of the possibility that it may react with sulfur dioxide to produce ammonium sulfate.[1]   Since sulfur dioxide is one of the principal air pollutants, it is of interest to determine the reaction of SO_2 in the atmosphere. The significance of ammonium sulfate in the atmosphere is related to health,[2] corrosion[3] and visibility.[4]

There are many methods available for the analysis of ammonium ion such as the Nessler,[5] indophenol,[6] ion selective electrode,[7] chemilumines- cence[8] and optical waveguide[9] methods. These methods appear to be adequate but suffer from either lack of sensitivity or selectivity, or are too complex and cumbersome to use.

This chapter demonstrates the simplicity and versatility of a relatively new technique called ion chromatography (IC) for the analysis of NH_4^+ in ambient aerosols.

Ion chromatography (IC) was first described by Small et al.[10]   Mulik et al.[11] applied IC to the analysis of sulfate and nitrate in ambient aero- sols and later applied IC to the analysis of atmospheric SO_2 described elsewhere in these proceedings.

The principle of ion chromatographic analysis of NH_4^+ and other cations is shown in Figure 4.1. Nitric acid is the eluent. The cations are sepa- rated in the analytical column, which contains a strong acid ion exchange resin. The separated cations pass into the suppressor column containing a strong basic ion exchange resin. The suppressor column exchanges its OH^- for the NO_3^- in the eluent converting the eluent into water while the sample cations pick up OH^- to form hydroxides. Thus, hydroxides enter the detector in a background of water and are easily measured with the conductivity cell.

Experimental

To evaluate the ion chromatograph as an instrument for the analysis of NH_4^+ in ambient aerosols, standard NH_4^+ concentrations were prepared from reagent grade ammonium sulfate. Detailed calibration curves were run periodically. For the data presented herein, however, the single-point external standard method was used. The sample is run first to estimate the species concentration; then a closely matched standard is assayed. Quantitation is obtained by the ratio of the peak area of the standard to the peak area of the sample, as shown in Equation 1.

$$A \quad = \frac{C_{Std}}{PA_{Std}} \quad x \quad PA_{Unk} \tag{1}$$

A = concentration of unknown $\mu g \ NH_4^+/ml$

C_{Std} = concentration of standard $\mu g \ NH_4^+/ml$

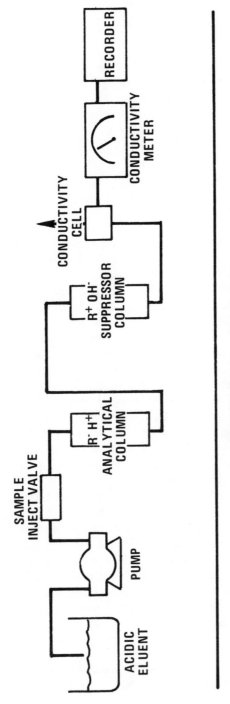

CATION ANALYSIS

SUPPRESSION COLUMN PRINCIPLE
ELUENT – HNO_3

$$H^+NO_3^- + R^+OH^- \longrightarrow R^+NO_3^- + H_2O$$

$$NH_4^+ NO_3^- + R^+ OH^- \longrightarrow NH_4^+OH^- + R^+NO_3^-$$

Figure 4.1 Ion chromatographic principle.

PA_{Std} = peak area of standard
PA_{Unk} = peak area of unknown

A Model 10 Ion Chromatograph (Dionex Corp.,* Palo Alto, California) was used for the analysis of ammonium ion in ambient aerosols. A schematic of the flow system is shown in Chapter 3, page 31. The flow system consists of a separator or analytical column, suppressor column, four solvent reservoirs, injection valve with 0.5-ml sample loop, two Milton Roy fluid pumps, conductivity detector and a valving system to direct the flow through various parts of the instrument. The system uses air-activated Teflon slider valves throughout.

RESULTS

A typical ion chromatogram of a standard solution of the cations of Na^+, NH_4^+ and K^+ is shown in Figure 4.2. Since it is possible that aliphatic amines could interfere with NH_4^+ assay, a standard solution of Na^+, NH_4^+ and K^+ was spiked with methylamine, dimethylamine and ethylamine to give the chromatogram shown in Figure 4.3. Although the resolution is not good for the aliphatic amines, Figure 4.3 shows that amines do not interfere with the NH_4^+ measurement. The eluent flow was also reduced from 60% to 30%, but this still did not resolve the K^+ from methylamine.

Figure 4.4 is an ion chromatogram of a Hi-Vol filter extract showing that it is possible to assay NH_4^+ in a preponderance of Na^+.

Excellent calibration curves were obtained at high NH_4^+ concentrations (Figure 4.5) and at much lower NH_4^+ concentrations (Figure 4.6).

Repeatability of IC measurement for NH_4^+ is shown in Table 4.1. The repeatability data reported in Table 4.1 were the average of 10 analysis at each concentration. The relative standard deviation was quite good, even at concentrations as low as 0.09 $\mu g/ml$.

An interlaboratory comparison of IC data on Hi-Vol filter extracts is shown in Table 4.2. Table 4.2 shows that Laboratory 1 performed IC analyses on Hi-Vol extracts on two different days and also performed colorimetric analysis on the same extracts. Laboratory 2 performed IC analyses on aliquots from the extracts. The excellent interlaboratory and intralaboratory agreement indicates that IC measurements of NH_4^+ are repeatable and that they agree with independent analyses techniques.

Table 4.3 shows a comparison of the IC method for NH_4^+ with the ion selective electrode on dichotomous filters that have been impregnated

---

*Any mention of commercial products does not constitute endorsement by the United States Environmental Protection Agency.

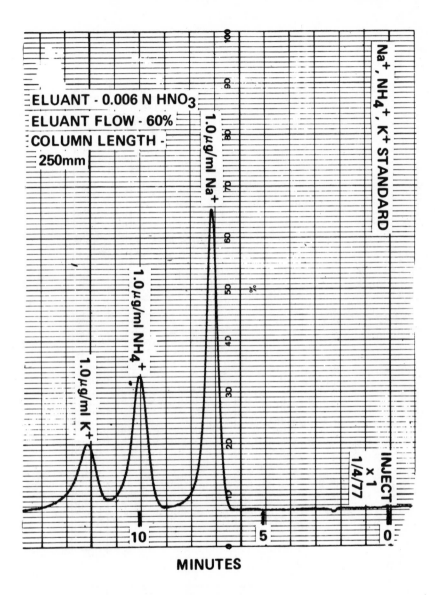

Figure 4.2 Ion chromatogram of Na^+, NH_4^+, K^+.

with NH_4^+ standards. The results of both methods appear to be low, indicating that a better extraction technique must be developed or that spiking was off. However, the agreement between the analytical methods is good.

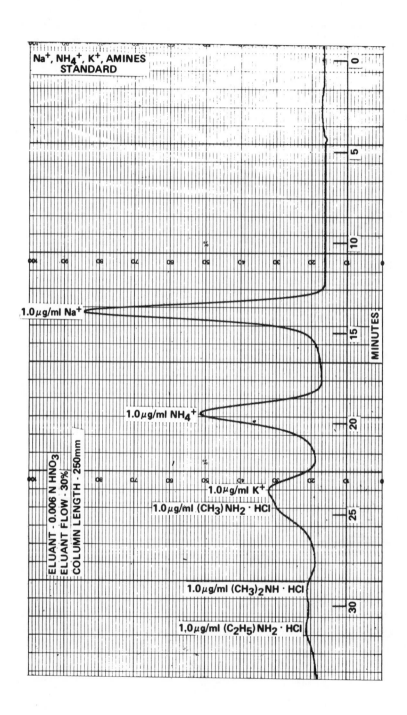

Figure 4.3 Ion chromatogram of Na^+, NH_4^+, K^+ and aliphatic amines.

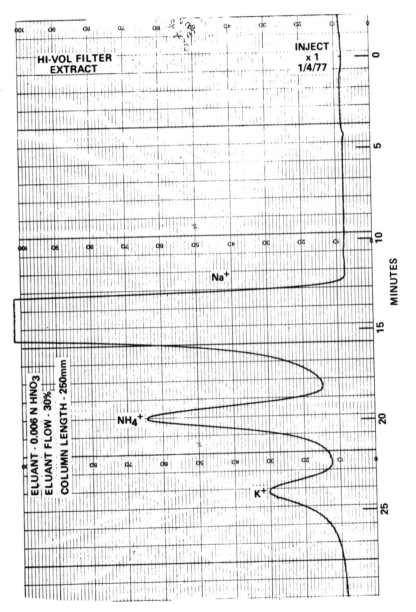

Figure 4.4   Ion chromatogram of Hi-Vol extract.

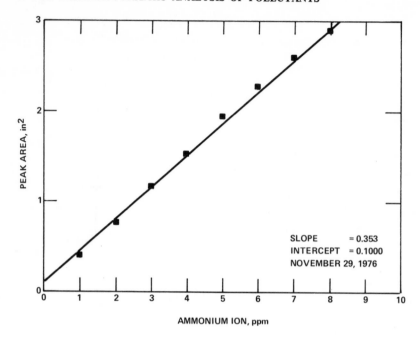

Figure 4.5   Calibration curve for NH_4^+ (0-10 ppm).

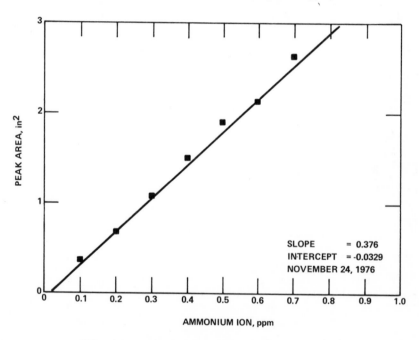

Figure 4.6   Calibration curve for NH_4^+ (0-1 ppm).

Table 4.1 Repeatability of IC Measurement for Ammonium Ion
(Average of 10 Determinations)

	9 μg/ml	0.9 μg/ml	0.09 μg/ml
Standard Deviation	0.02	0.02	0.03
Relative Standard Deviation	0.9	1.8	2.9

Table 4.2 Interlaboratory Comparison of Ion Chromatographic Method for NH_4^+

	NH_4^+ (μg/ml)			
	Lab 1	Lab 1	Lab 1	Lab 2
Hi-Vol Filter Sample No.	Colorimetric 2/14/77	IC 2/14/77	IC 2/17/77	IC 2/17/77
ATVR	4.3	4.1	4.1	4.4
ATVS	3.9	3.7	3.8	4.1
ATWX	6.8	6.5	6.8	6.5
ATXB	5.0	4.7	4.9	5.0
ATXE	7.9	7.9	8.0	8.3
ATXW	9.8	10.0	10.0	9.5
ATYD	4.8	4.6	4.8	4.8
ATYE	9.3	9.1	9.2	9.1
ATYF	2.2	2.3	2.2	2.4

Table 4.3 $(NH_4)_2SO_4$ Standards Impregnated on 37-mm Teflon Dichotomous Filters

		Found (μg)	
Sample No.	Added (μg)	IC	Ion Selective Electrode
1	87.0	77	72
2	82.0	74	72
3	82.1	72	71
4	60.2	66	65
5	80.3	67	62
6	71.5	65	64
7	101.0	92	91
8	145.4	132	128

Table 4.4 shows a comparison of IC with the ion selective electrode on actual field samples obtained from West Virginia with a dichotomous

Table 4.4 Comparison of Ion Chromatographic Method with Ion Selective Method on Actual Field Samples from West Virginia

Dichotomous Filter Sample No.	NH_4^+ ($\mu g/ml$)	
	Ion Selective	Ion Chromatography
11	0.58	0.61
21	0.45	0.39
78	0.49	0.53
80	0.98	1.04
191	0.82	0.77
193	0.69	0.65
209	1.16	1.04
15	0.36	0.27
84	0.28	0.20
189	0.34	0.24

sampler. There appears to be good agreement between the two techniques, except at the very low concentrations.

Preliminary experiments have also shown that gaseous NH_3 can be collected in a 3% oxalic acid solution[12] or on a filter coated with oxalic acid and analyzed with the ion chromatograph.

CONCLUSIONS

Ion chromatography has the sensitivity, selectivity, precision and accuracy to assay for NH_4^+ in ambient aerosols obtained either from Hi-Vol filters or from dichotomous filters.

REFERENCES

1. McKay, H. A. C. "Ammonia and Air Pollution," *Chem. Ind.* 1162 (1969).
2. Amdur, M. O. and M. Corn. "The Irritant Potency of Zinc Ammonium Sulfate of Different Particle Dyes," *Am. Ind. Hyg. Assoc. J.* 24:326 (1963).
3. Mimer, S. "Air Pollution Aspects of Ammonia," National Air Pollution Control Association, Consumer Relations and Environmental Health Services, Department of Health, Education and Welfare, USA (1969).
4. Eggleton, A. E. "Chemical Composition of Atmospheric Aerosols on Tees-side and its Relation to Visibility," *Atmos. Environ.* 3:355 (1969).

5. Stern, A. C. "Analysis, Monitoring and Surveying," *Air Pollution,* Vol. II (New York: Academic Press, 1968).
6. Rommers, P. J. and J. Visser. "Spectrophotometric Determination of Micro Amounts of Nitrogen as Indophenol," *Analyst* 94:653 (1969).
7. Eagan, M. L. and L. Dubois. "Determination of Ammonium Ion in Airborne Particulates With Selective Electrodes," *Anal. Chim. Acta* 70:157 (1974).
8. Hodgeson, J. A., K. A. Rehme, B. A. Martin and R. R. Stevens. "Measurements for Atmospheric Oxides of Nitrogen and NH_3 by Chemiluminescence," presented at 1972 APCA Meeting, Miami, Florida, June 1972.
9. David, D., M. Willson and P. Ruffin. "Direct Measurement of NH_3 in Ambient Air," *Anal. Letters* 9(4):389 (1976).
10. Small, H., T. Stevens and W. Bauman. "Novel Ion Exchange Chromatographic Method Using Conductimetric Detection," *Anal. Chem.* 47:1801 (1975).
11. Mulik, J., R. Puckett, D. Williams and E. Sawicki. "Analysis of Nitrate and Sulfate in Ambient Aerosols," *Anal. Letters* 9(7):653 (1976).
12. Shendrikar, A. and J. Lodge, Jr. "Micro Determination of Ammonia by the Ring Oven Technique and its Application to Air Pollution Studies," *Atmos. Environ.* 9 (1975).

5

PRACTICAL EXPERIENCE ON THE USE OF ION CHROMATOGRAPHY FOR DETERMINATION OF ANIONS IN FILTER CATCH SAMPLES

J. Lathouse and R. W. Coutant

Battelle-Columbus Laboratories
Columbus, Ohio 43201

ABSTRACT

The IC has been utilized at Battelle-Columbus for determination of anions in filter catch environmental samples for a period of about half a year (as of January 1977). We will discuss the pragmatics of obtaining useful, validated data in terms of procedure, problems, precision, cross check data and area vs peak height.

Cross check data include comparison between IC data for $SO_4^=$ and NO_3^- and barium perchlorate titration (thorin indicator) for $SO_4^=$ and brucine colorimetric method for NO_3^-.

INTRODUCTION

The analytical labs at BCL perform a dual role, providing analytical services for other sections and departments as well as functioning as a part of our own research efforts. Typical samples range in origin from ambient particulate to source effluents to process stream products. We therefore must deal with a large volume of diverse samples, and are continually seeking analytical methods with improved versatility and speed without sacrificing sensitivity and accuracy. We were quite interested in the work of Small et al.[1] and, after hearing the preliminary report of Mulik,[2] obtained a Dionex Model 10 Ion Chromatograph for our laboratory.

In anticipation of receipt of this instrument, we retained a number of fluoride, nitrate and sulfate samples that had already been analyzed by conventional wet chemistry methods. The purpose of this chapter is to present comparison of results obtained on these samples by IC and wet chemical procedures, and to discuss some of our experience relative to practical application of the IC method. In particular, we will address the effects of column aging on apparent performance of the IC.

PROCEDURES

In general, the preparation of samples for the IC is governed by the amount of sample available, the type of sample being analyzed and the expected concentrations of species of interest.

Filter samples that are readily wettable are cut up and leached with deionized water on a steam bath for a period of about two hours. These samples are then cooled and filtered using 0.22 μ Millipore filters. Filters that do not wet readily are cut and leached using glass rods to hold the pieces under water. After filtration, all samples are made up to volume, with aliquots being taken for analysis.

Blank filters are run with each set of samples, and all apparatus are rinsed with water between samples. The procedure for the preparation of filter samples was used by Blosser et al.[3] The eluent used throughout this work was 0.003 M NaHCO$_3$ and 0.0024 M Na$_2$CO$_3$. Pump rate was 35%, and a 0.1-ml sample loop was employed.

RESULTS

Method Comparisons

Although ion chromatography shows promise for analysis of many different anions and cations, our use thus far has been restricted to determinations of sulfate, nitrate, fluoride and chloride samples, with primary emphasis on sulfate and nitrate. Comparative data have been obtained for fluoride, sulfate and nitrate. These data were generated from round robin tests as well as from comparison between IC and other methods of analyses.

Round Robin Analyses

We participated in a round robin test for SO$_4^=$ and NO$_3^-$ determination on strips of glass fiber filter. While the final results, which involve many analytical techniques, are not yet published, we feel our results demonstrate the validity of the extraction procedure as well as the validity of IC at least on these round robin samples, which were part of an EPA performance audit.

Table 5.1 gives the results obtained as well as the "sample range" and "target range." The sample range indicates the best that can be expected under conditions existing during the audit and the target range assumes the greatest amount of variability that can occur. Thus, results within the sample range suggest any errors that are not detectable, and results within the sample range suggest whether the analytical process was operating properly. As can be seen from the data, the agreement was excellent except for one nitrate determination which, after the fact, was found to be derived from a chromatogram with a baseline shift.

Table 5.1 Analyses of Round Robin Samples

Sample	Reported Value	Sample Range	Target Range
$SO_4^=$ $\mu g/m^3$			
1	22.1	20.1 -22.2	18.0 -24.3
2	20.8	20.1 -22.2	18.0 -24.3
3	15.5	14.8 -16.4	13.3 -18.0
4	15.8	14.8 -16.4	13.3 -18.0
5	25.2	23.6 -26.1	21.1 -28.6
6	4.50	4.25- 4.69	3.80- 5.14
NO_3^- g/m^3			
1	11.8	10.4 -11.5	9.28-12.6
2	11.1	10.4 -11.5	9.28-12.6
3	5.25	4.91 5.43	4.40- 5.95
4	5.42	4.91- 5.43	4.40- 5.95
5	2.25[a]	1.84- 2.04	1.65- 2.23
6	7.65	7.01- 7.75	6.27- 8.49

[a]In the NO_3^- results sample #5 is out of the target range. A review of the sample scan shows a shifted baseline of 0.5 unit. The results would then be 2.14. This correction apparently would have given better results.

Fluorides

The fluoride samples examined were filter catches of welding fumes. In this case, total fluoride analysis was desired, and samples were fused prior to leaching. The classical Willard-Winkler distillation procedure was used, followed by thorium nitrate colorimetric titration in alizarin red as the indicator for the first analyses.

The sample-prepared solutions were used for the IC analyses. Table 5.2 shows the sample preparation for fluoride analyses. A 0.1-ml loop was used for the IC samples; the results are shown in Table 5.3. It

Table 5.2 Preparation of Fluoride Samples

Sample	Filter Size	Amount Used	Amount Distilled	Colorimetric Titration Tap	IC	Sensitivity
1-10	325 mm	1/4	500 ml	0.5 ml	0.1 ml	X100

Table 5.3 Comparative Fluoride Analyses

Sample	Total Fluoride (mg)	
	Colorimetric Titration	IC
1	34	37
2	42.4	42
3	38.8	38.75
4	43.2	45.25
5	45.2	47
6	46.0	49.75
7	19.2	21.0
8	10.8	12.0
9	32.0	36.5
10	28.4	33.5
11	48.8	53.0

can be seen from this table that agreement between the two methods is reasonably good with the IC result tending to be about 7% higher than those obtained by titration. These were among the first samples analyzed by IC; more recent results tend to be in slightly better agreement as experience has been gained in applying the method.

Nitrates

Nitrate samples examined were ambient particulate catches that had been spiked with nitric acid in some cases. The brucine method was used as a reference for these samples. The preparation data are given in Table 5.4. Data for these samples are given in Table 5.5. For the most part, these results again indicate good agreement between the reference method and IC. The IC results tend to be high by about 4%, and the standard deviation between the results of the two methods is 10%.

Table 5.4 Preparation of Nitrate Samples[a]

Sample	Amount Taken	Dilution (ml)	Brucine Method Tap		IC Sensitivity
1	1/2	10	5 ml		X 3
2	1/2	10	5 ml		X 10
3	1/2	10	5 ml		X 10
4	1/2	10	5 ml		X 3
5	1/2	10	5 ml		X 3
6	1/2	10	5		X 3
7	1/2	10	1		X 3
8	1/2	10	1		X 10
9	1/2	10	1		X 10
10	1/2	10	5		X 3
11	1/2	10	5		X 3
12	1/2	10	5		X 3
13	1/2	10	5		X 3
14	1/2	10	5		X 3
15	1/2	10	5		X 3
				Redilution	
16	1/4	50	0.5	-	X 10
17	1/4	50	0.1	1/25	X 3
18	1/2	10	0.5	-	X 10
19	1/4	50	0.5	1/25	X 3

[a]Filter size is 47 mm. Sample loop is 0.1 ml.

Sulfates

The sulfate samples examined were probe washings taken from a Method 5 train used for source sampling. Table 5.6 gives data for sample preparation for sulfate determination. Table 5.7 shows a comparison of IC results on these samples and corresponding results obtained by titration with barium perchlorate, using thorin as the indicator. With this set of data, the IC results tended to be low by about 2% (average deviation) and the standard deviation is 6.5%.

Effect of Column Aging

Sensitivity

Normal procedure in our laboratory calls for running a new set of standards every day. Over an extended period of use, we have noted that the retention time for a given anion and the apparent sensitivity change considerably. For example, Figure 5.1 shows several sulfate

Table 5.5   Comparative Nitrate Analyses

| Sample | Total NO$_3^-$ (μg) | |
	Brucine	IC
1	37	38.5
2	45	47
3	117	140
4	6	5
5	11	10
6	15	20
7	190	211
8	122.5	145
9	342.5	390
10	24	25
11	46	45
12	3.6	2.5
13	18.8	20.0
14	58.2	65
15	240	245
16	1,060	1,060
17	22,000	20,870
18	680	750
19	15,600/16,500	15,250

Table 5.6   Preparation of Sulfate Samples[a]

Sample	Weight (g)	Tap for Titration Volume (ml)	Sensitivity
1	0.1627	5	X 10
2	0.1568	5	X 100
3	0.1727	5	X 30
4	0.1702	5	X 10
5	0.1131	5	X 30
6	0.0810	5	X 30
7	0.1594	5	X 100
8	0.1667	5	X 100
9	0.1531	5	X 100
10	0.1559	5	X 100
11	0.1618	5	X 100
12	0.1713	5	X 100
13	0.0567	5	X 30
14	0.1490	5	X 1000 X 300
15	0.0516	5	X 100

[a]For all samples Ba(ClO$_4$)$_2$ dilution was 50, the tap for titration was 5 ml and the sample loop was 0.1 ml.

Table 5.7   Comparative Sulfate Analyses

Sample	Percentage $SO_4^=$	
	Ba(ClO$_4$) Titration	IC
1	1.34	1.19
2	9.98	9.66
3	1.73	1.48
4	0.965	0.960
5	3.20	3.25
6	3.34	3.09
7	8.27	8.55
8	10.72	11.47
9	5.24	5.23
10	9.85	10.94
11	6.75	6.95
12	10.88	11.46
13	3.75	3.43
14	25.21	31.46/32.09
15	8.42	7.27

calibration curves obtained for one column over a four-month period. During this time, 745 samples (excluding standards) were run through the column, and the operating parameters were constant throughout. On September 3, 1976, a full scan was completed in 12 minutes; on December 16, 1976, the corresponding scan time was 6 to 7 minutes.

It is obvious from Figure 5.1 that the apparent sensitivity of the instrument also changed during this time. Although the data in Figure 5.1 are based on measurements of peak height, comparison of peak areas shows a similar trend. Figure 5.2 shows a plot of sensitivity (height/μg SO$_4$) vs the number of samples passed through the instrument. It can be seen from this plot that the apparent sensitivity does correlate well with column usage.

These results suggest that irreversible sorptions of sulfate occur to some extent on the fresh column, but we have not yet tested this hypothesis directly.

Column Lifetime

As noted above, retention times gradually decrease with column usage until the chromatograms finally show little or no resolution of the ion peaks of interest. With our first column, only 263 samples could be run before performance became unacceptable. With the second column, we were able to run 834 samples. Neither of these columns was used with a

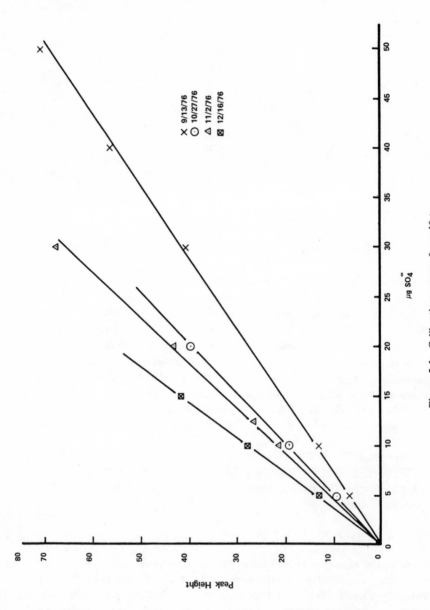

Figure 5.1 Calibration curves for sulfate.

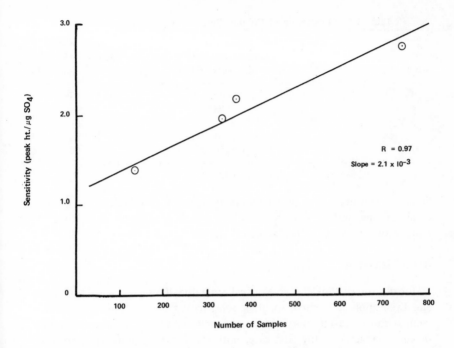

Figure 5.2 Effect of usage on apparent sensitivity.

precleaning column. Currently, we are using a precleaning column, and have run 220 samples (February 18, 1977). Hopefully, the use of the precleaning column and improvements made in the resin will extend the lifetime of the columns appreciably.

Changing Columns

Because of the effects of column aging on apparent sensitivity and retention time, it is to be expected that changing from one column to the next will result in different absolute responses, but the analytical results for the same samples put through different columns might be expected to be identical. Table 5.8 shows a comparison of results obtained for five filter blanks analyzed for sulfate using our second and third columns. In this single example, the new column gave results that averaged 6.4% higher than those obtained with the old column.

Effect of Temperature

Most of our measurements have been conducted at $25 \pm 1°C$, so that our data concerning the effect of temperature on IC response are very

Table 5.8   Comparison of Column Response with Identical Samples

Sample	Column II January 14, 1977 mg SO_4 Total Filter	Column III January 18, 1977 mg SO_4 Total Filter
3	0.860	0.935
4	1.0105	0.990
5	0.925	1.015
6	0.9505	0.985
7	0.895	1.00

limited.   However, as a casual observation, we normally expect a change of about one unit of peak height (10 x scale) per degree, with higher temperature yielding higher apparent response.

Range Selection

Ideally, one should be able to calibrate the IC on one detector range, and then apply this curve to other sensitivity ranges.   Table 5.9 shows a comparison of results obtained on the same samples with two different detector ranges.   In the first case, both standards and samples were run on the 30x scale, in the second case, the standards and samples were run on the 100x scale, and in the third case, the standards were run on the 100x scale, but the samples were run on the 30x scale.   For all these samples, results obtained on 100x were higher than those obtained on 30x, with the average difference of the calculated results being 12%.

Table 5.9   Total Sulfate Found, mg

Samples	Standards & Samples x30	Standards & Samples x100	Samples Run x30 Read on x100 Curve and Calculated
A	9.58	10.3	8.88
B	12.03	12.3	11.1
C	10.6	11.0	9.81
D	11.55	11.8	10.61
E	12.10	12.25	11.21
F	9.03	9.75	8.37
G	9.7	10.30	9.50

DISCUSSION AND CONCLUSIONS

In general, we have been pleased with the performance of the IC as a tool for rapid analysis of diverse types of samples, and it is particularly useful for analysis of more than one anion per sample. However, there are some precautions that need to be taken in routine application of the method. The effect of column aging appears to be due to gradual degeneration of the resin and requires standards to be run frequently. It also is recommended that new standards be run whenever particular samples require the use of a different detector range. In comparison of the IC results for sulfate, nitrate and fluoride analyses with those obtained by titration procedures, deviations are usually less than 10% and are generally considered acceptable. Inasmuch as the results obtained by titration procedures are often highly dependent on the experience and judgment of the analyst, we feel that the IC results are preferable for routine analytical work.

REFERENCES

1. Small, H., T. S. Stevens and W. C. Bauman. *Anal. Chem.* 47:1081 (1975).
2. Mulik, J.D., R. Puckett, D. Williams and E. Sawicki. Preliminary Report given at EPA, March 1976.
3. Blosser, E. R., L. J. Hillenbrand and J. Lathouse. NBS Special Publication 422, *Sampling and Analysis for Sulfur Compounds in Automobile Exhaust* Volume I (August 1976), p. 389.

DISCUSSION

What happens to the retention times and resolutions as the columns are used over and over again?

Your retention times shorten. For example, sulfate that originally appears in 10 minutes, with much use appears at 7 or 6 minutes. Also, the peak sharpens. I have the very same data for the fast column that we ran only 263 samples through. You get the same type of response.

What type of column do you use?

It is the short, 3 x 100 mm column that Dionex sells. It has the same resin in it as the large 3 x 500-mm analytical column.

After you had used your columns for a very long time, did you try to clean them up?

Yes we did, for 24 hours with strong carbonate, and it did not help at all.

What happens to the columns over a long period of use?

It is my opinion that there is a gradual poisoning of the column.

Was the same sample analyzed by the titration and the ion chromatographic methods?

The same sample was analyzed by the titration and ion chromatographic methods.

How can you tell when your column is not working satisfactorily?

You can tell soon enough. You come in some morning and get no separation at all between your sulfate and nitrate and fluoride and chloride. This difficulty is easy to spot. The bands just sort of lump together. When this happens, you know your column is not operating properly.

After our first column went out so quickly after fewer than 300 samples, we did send it back to Dionex. I don't know whether Mr. Smith found out anything about the resin or not. Did you?

This is a problem we are still working on. We have answers, which we will report in the near future.

6

ANALYSIS OF AIR PARTICULATES BY ION CHROMATOGRAPHY: COMPARISON WITH ACCEPTED METHODS

F. E. Butler, R. H. Jungers, L. F. Porter, A. E. Riley and F. J. Toth

Analytical Chemistry Branch
Environmental Monitoring and Support Laboratory
Environmental Protection Agency
Research Triangle Park, North Carolina 27711

ABSTRACT

Ion chromatographic analysis is applied for the determination of anions and cations solubilized from Hi-Vol fiberglass filters. Two Dionex ion chromatographs were used to determine background levels of ions in 600 unexposed or "blank" filters within a three-week period. Exposed filters from the National Air Monitoring Network, Los Angeles Catalyst Study, and Continuous Health Air Monitoring Program were analyzed. Precipitation samples were also analyzed by ion chromatography. Results are compared to those obtained by standard autoanalyzer and ion specific electrode techniques. Factors considered are time required per analysis, sensitivity, accuracy and versatility. These data and quality control results are discussed.

INTRODUCTION

This laboratory is responsible for the determination of anions and cations in support of routine and special projects concerning environmental air purity. Analyses are performed on either dissolved particulates from fiberglass and membrane filters or on precipitation samples. In excess of 20,000 determinations are made each year for the concentration of sulfate, nitrate, ammonium, fluoride and chloride ions in these samples. Greater than 90% of anion analyses are for sulfate and nitrate ions. With the advent of commercially available ion chromatographs, we were interested in comparing results obtained by these instruments with results from currently used techniques. Mulik et al.[1] recently reported the successful application of ion chromatography for the analysis of sulfate and nitrate in ambient aerosols.

The purpose of this chapter is to report the preliminary results of tests for sensitivity, precision, accuracy and rapidity of ion chromatography and to compare these results with those obtained by automated colorimetric and ion specific methods currently used.

EXPERIMENTAL

Ion Chromatographs

Two Model 10 ion chromatograph instruments were purchased from the Dionex Corp.,* Sunnyvale, California. Each instrument is equipped with both anion and cation analytical and suppressor columns. Sample loops are 200 mm (approximately 100-μl capacity).

Optimum Anion Peak Separations

Both the eluent solution concentration and pump pressures were varied to determine the best conditions for rapid analysis of anions. Various concentrations of sodium carbonate and sodium bicarbonate were used. Although more rapid elution times were possible with some combinations of the eluents for certain anions, the best mixture for all seven anions tested was that recommended by the vendor. This mixture, 0.003 M NaHCO$_3$ and 0.0024 M Na$_2$CO$_3$, was prepared in 20-liter batches.

The higher pressures, of course, gave more rapid separations. The vendor recommends a maximum of 550 psi for these low-pressure systems. Our systems were therefore operated at 530 to 550 psi peak pressure (4 ml/min).

Figure 6.1 shows separations for one instrument. The total time from injection (pip) to return to baseline after the last ion (sulfate) was 11.5 minutes. Using the same eluent solution and pressure, the other instrument required 14 minutes for complete elution.

---

*Any mention of commercial products does not constitute endorsement by the United States Environmental Protection Agency.

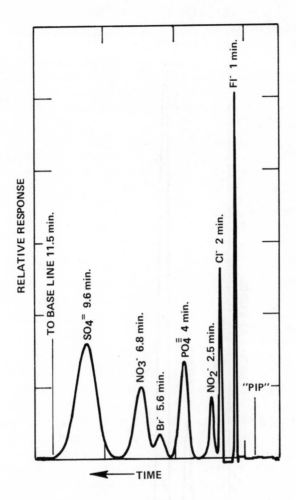

RELATIVE RESPONSE

TO BASE LINE 11.5 min.

$SO_4^=$ 9.6 min.

NO_3^- 6.8 min.

Br^- 5.6 min.

$PO_4^{\equiv}$ 4 min.

NO_2^- 2.5 min.

Cl^- 2 min.

Fl^- 1 min.

"PIP"

◄——TIME

Figure 6.1 Anion spectra.

Separation of Cations

Cations were eluted with 0.005 M nitric acid at 400 psi. Figure 6.2 shows the separation of sodium, ammonium and potassium ions. The complete elution for these three cations requires 10 minutes.

Regeneration of Columns

The analysis of anions is the major topic of this chapter. During one three-week period, 600 samples were analyzed (see "Blank" Filters, page 72). At the conditions described above, regeneration with the

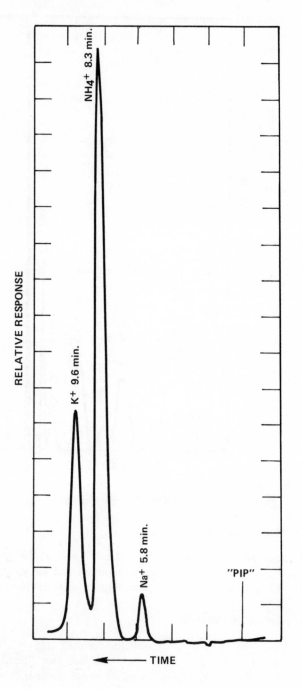

Figure 6.2   Cation spectra.

recommended 1 N sulfuric acid was required after 5 to 6 hours of continuous operation. Regeneration and rinse times were 25 and 15 minutes, respectively.

Precision

Prior to analysis, each instrument was calibrated daily with several solutions of mixed anion standards. After every tenth sample, one mixed standard solution was injected. Results for one of these solutions are shown in Table 6.1. During a two-week period this solution was injected 29 times. Note that the relative standard deviations for most of the anions are in the range 5 to 10%. The worst precision was obtained for chloride. Although all operating parameters were maintained as constant as possible, better precision would probably be achieved if all data were gathered in a shorter period of time.

Table 6.1 Precision for Mixed Anion Standards Injected 29 Times During a Two-Week Period

	Fl^-	Cl^-	NO_2^-	$PO_4^{\equiv}$	Br^-	NO_3^-	$SO_4^=$
$\mu g/ml$	0.25	0.50	0.50	2.50	0.50	1.50	2.50
Mean Scale Reading	2.69	1.75	0.46	0.68	0.19	0.53	0.74
Standard Deviation	0.28	0.74	0.11	0.05	0.03	0.05	0.04
Maximum	3.36	4.27	0.74	0.78	0.27	0.64	0.89
Minimum	2.10	0.63	0.30	0.56	0.14	0.43	0.67

The response for each ion is obtained over a wide range of concentrations, as shown in Figure 6.3. This figure shows the computer printout of all data for sulfate ion concentration between 0.50 and 8.75 $\mu g/ml$ The best straight line fit for these points does not quite intercept the origin. In practice, all unknown samples analyzed during the two-week period were calculated using the linear regression formula shown in the figure. Peak height comparisons were made for standards and samples.

Figure 6.4 contains the data for 2.5 $\mu g/ml$ sulfate shown in Figure 6.3. The distribution around the mean scale reading of 0.74 is a normal distribution.

Sensitivity

The mixed standard solution containing the seven anions was serially diluted to determine the best practical sensitivity for each ion. The criterion used was peak responses equal to twice the background

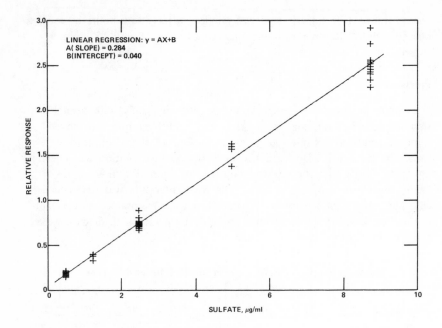

Figure 6.3   Sulfate calibration curve.

fluctuations, generally at a scale reading of one micromho. These sensitivities are contained in Table 6.2. Note that with the exception of the chloride ion, most ions were detected at approximately 0.1 μg/ml levels.

RESULTS AND DISCUSSION

Methods Used for Data Comparisons

Several types of samples that contained different concentrations of anions and cations were analyzed by both the ion chromatography method and methods used in this laboratory.

Sulfate, nitrate and ammonium ions are determined using Auto-Technicon II procedures.[2]   Sulfate is analyzed by the methylthymol-blue (MTB) method.   The reaction of sulfate with MTB-barium complex results in free ligand, which is measured colorimetrically; nitrate is reduced to nitrite which reacts with sulfanilamide to form a diazo compound.   This is then reacted to an azo dye for colorimetric measurement.   Ammonium ions are reacted with sodium phenolate and sodium hypochlorite to produce a blue-colored complex for colorimetric measurement.

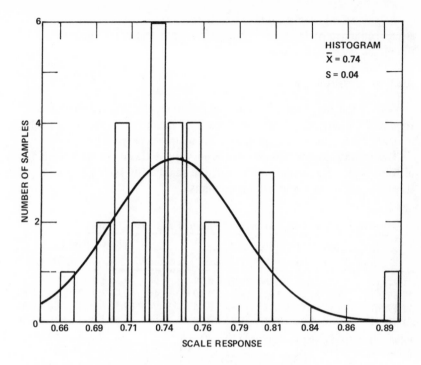

Figure 6.4   Histogram for 2.5 μg/ml sulfate.

Table 6.2 Maximum Sensitivity for Anions, μg/ml

Fl⁻	Cl⁻	NO₂⁻	PO₄³⁻	Br⁻	NO₃⁻	SO₄²⁻
0.008	0.4	0.07	0.10	0.12	0.14	0.16

Chloride ions are analyzed using a Technicon I Autoanalyzer. The chloride ion is reacted with mercuric thiocyanate and ferric ammonium sulfate. The resultant ferric thiocyanate is determined colorimetrically.

Solutions containing fluoride are buffered and the fluoride is determined by an ion specific electrode technique.[3]

Sample Preparation

Ions in fiberglass filters were solubilized by soaking 3/4-in. x 8-in. strips of filters in 50 ml of deionized water. The samples were then placed in an ultrasonic bath and then filtered.[2] Aliquots of each sample solution were analyzed by both the current methods described above and by ion chromatography. Samples were prepared in groups of 100. Each group contained at least seven known standard and blank quality control filters.

Analysis of "Blank" Filters

One of the first uses of the ion chromatography method in this laboratory was the analysis of 600 unexposed or blank filters which represented 300,000 filters purchased for use in air sampling programs.

The quality control samples analyzed with these filters are summarized in Table 6.3. The actual results and the number of standards run are shown in the nitrate and sulfate columns.

Table 6.3   Analysis of Standard and Blank Samples

Nitrate (μg/ml)			Sulfate (μg/ml)		
Actual	Dionex	Technicon	Actual	Dionex	Technicon
31.93 (13)	31.05 ± 1.45	29.92 ± 0.79	9.96	9.54 ± 0.77	9.60 ± 0.24
17.88 (6)	17.24 ± 0.63	16.47 ± 0.59	53.40	50.69 ± 2.33	49.38 ± 1.80
11.49 (6)	10.88 ± 0.38	10.90 ± 0.45	34.74	32.39 ± 0.74	33.00 ± 0.58
1.81 (4)	2.20 ± 0.65	1.89 ± 0.08	4.00	4.53 ± 0.52	4.50 ± 0.60
Filter "Blank" (19) 0.01	0.22			0.07	0.39
Water "Blank" (7) <0.01	0.03			<0.01	0.01

Aliquots of each solution were analyzed in both the Dionex and Technicon instruments. Note that the mean results for methods are similar. The variations in results by Dionex are higher.

One advantage of the Dionex method was demonstrated with these control samples. The autoanalyzers were calibrated from 0 to 10 μg/ml for sulfate ion and 0 to 5 μg/ml for nitrate ion. Standard samples assaying higher than these levels required appropriate dilution and reanalysis in the autoanalyzer. The Dionex required only a switchover to a higher attenuation, provided the analyst were present to observe the elution of off-scale peaks.

Table 6.4 shows typical fluoride, chloride and phosphate ion concentrations in the 600 blank filters. The Dionex results can be compared with ion specific electrode results for fluoride and with Technicon I results for chloride. The filter preparation by the vendor included either a phosphoric acid or a hydrochloric acid wash step. This is indicated in the phosphate and chloride ion results—samples 1 to 10 showing high phosphate and low chloride, and samples 45 to 54 showing the reverse.

Table 6.4   Analysis of Fluoride and Chloride

Sample[a]	Fluoride		Chloride		Phosphate
	Ion Specific Electrode (μg/ml)	Dionex (μg/ml)	Technicon (μg/ml)	Dionex (μg/ml)	Dionex (μg/ml)
1	0.29	0.28	0.2	0.1	5.7
2	0.28	0.26	0.1	0.2	5.1
3	0.73	0.72	0.1	0.1	4.0
4	0.81	0.78	0.1	0.1	4.4
5	0.69	0.64	0.1	0.1	4.0
6	0.52	0.48	0.2	0.1	4.3
7	0.54	0.44	0.1	0.1	3.8
8	0.49	0.46	0.1	0.1	4.3
9	0.73	0.61	0.2	0.1	4.0
10	0.46	0.24	0.1	0.1	5.2
45	0.57	0.62	4.1	4.3	<0.1
46	0.56	0.56	3.4	3.8	<0.1
47	0.54	0.54	3.1	3.6	<0.1
48	0.56	0.56	3.8	3.6	<0.1
49	0.58	0.60	3.7	4.0	<0.1
50	0.57	0.59	4.1	4.8	<0.1
51	0.57	0.59	3.6	3.3	<0.1
52	0.52	0.56	2.7	3.0	<0.1
53	0.56	0.57	2.7	2.6	<0.1
54	0.57	0.57	2.3	2.5	<0.1

[a]Fiberglass samples 1 to 10 washed with phosphoric acid; fiberglass samples 45 to 54 washed with hydrochloric acid.

The presence of phosphate ion interfered with the Technicon analysis for sulfate by producing high results and erratic baseline readings. This problem has since been alleviated.

NASN Samples

Samples from the National Air Surveillance Network were selected for comparative analysis. These results are presented in Table 6.5. Nitrate and sulfate ions were determined by the Technicon method by EPA and/ or Northrop Services Inc. (NSI) laboratories for comparison with the Dionex results. Results over a wide range of concentrations showed good agreement.

Table 6.5   Analysis of NASN Samples

| | Nitrate (μg/ml) | | Sulfate (μg/ml) | | |
| | Technicon | | Technicon | | |
Sample	NSI	Dionex	NSI	EPA	Dionex
1	14.7	14.4	40.9	40.9	43.2
7	5.3	5.4	24.2	23.8	22.5
11	3.5	4.0	110.8	103.6	119.7
17	9.8	9.5	202.8	198.0	216.0
24	0.6	0.8	113.6	106.8	112.7
36	28.0	25.4	43.2	43.0	44.4
38	12.3	11.7	73.9	72.8	77.7
67	16.9	15.5	99.5	99.2	105.6
70	18.9	17.3	34.7	34.2	33.3
73	9.4	8.8	56.7	56.5	56.5
86	13.1	13.0	33.8	33.2	31.8
96	24.9	22.1	50.6	50.4	53.3

Precipitation Samples

Table 6.6 shows analysis of precipitation samples by the methods. The concentrations of nitrate and sulfate ions are much lower than in NASN samples. Results may be compared with National Bureau of Standard samples in this table.

Table 6.6   Analysis of Precipitation Samples

| | Nitrate (μg/ml) | | | Sulfate (μg/ml) | | |
Sample	Technicon	Dionex	Actual	Technicon	Dionex	Actual
1	1.7	2.1	-	3.5	3.5	-
2	0.4	0.5	-	1.4	1.6	-
3	1.4	1.3	-	3.0	3.0	-
4	0.6	0.7	-	1.1	1.4	-
5	0.0	0.1	-	0.3	0.5	-
6	2.1	2.0	-	9.4	9.1	-
7	1.3	1.2	-	5.7	5.6	-
8	2.9	2.8	-	6.9	6.9	-
NBS	0.2	0.2	0.11	1.0	1.2	1.08
NBS	1.1	0.9	0.66	3.6	3.5	3.56
NBS	0.2	0.2	0.11	1.0	1.1	1.08
NBS	11.0	10.9	10.88	9.7	9.8	10.00

Nitrate and Nitrite Results

The Technicon method is based on reduction of the nitrate ion to nitrite prior to azo colorimetric analysis. Table 6.7 shows comparisons of low-level samples, which contained both nitrate and nitrite ions. By the Technicon method one result was obtained and compared with the sum of the nitrate and nitrite assays obtained by ion chromatography. Note the excellent agreement.

Table 6.7 Comparison of Nitrite Plus Nitrate by Dionex vs Combined NO_2 Plus NO_3 by Technicon

Sample	Dionex (μg/ml)			Technicon (μg/ml)
	NO_2^-	NO_3^-	Total	$NO_2 + NO_3$
38	0.38	0.32	0.7	0.7
39	0.52	1.47	2.0	2.0
40	0.34	1.67	2.0	2.0
41	0.79	2.48	3.3	3.2
42	0.58	0.52	1.1	0.7
43	0.55	5.04	5.6	5.8
44	0.58	1.79	2.4	2.1
45	0.69	2.93	3.6	3.5
46	0.69	0.40	1.1	0.7
47	0.62	2.88	3.5	3.1
48	0.65	2.36	3.0	3.0
49	0.34	1.67	2.0	2.2
50	0.21	0.32	0.5	0.6
51	0.41	2.36	2.8	2.8
52	0.76	3.60	4.4	4.3
53	0.76	2.65	3.4	3.2
54	0.86	1.70	2.6	2.1
55	0.45	1.44	1.9	1.8
56	0.45	2.30	2.8	2.8

Analysis of Cations in NASN Samples

Ammonium, sodium and potassium cations found in NASN filters by the different methods are shown in Table 6.8. Once again, there was excellent agreement for ammonium analysis between the Technicon and Dionex methods.

Table 6.8 Analysis of Cations in NASN Samples

| | Ammonium Ion | | | Sodium | Potassium |
| | Technicon | | | Dionex | Dionex |
Sample	NSI (μg/ml)	EPA (μg/ml)	Dionex (μg/ml)	(μg/ml)	(μg/ml)
ATVR	4.3	4.1	4.1	19.7	2.6
ATVS	3.9	3.7	3.8	19.0	1.4
ATWX	6.8	6.5	6.8	20.4	1.6
ATXB	5.0	4.7	4.9	13.5	1.0
ATXE	7.9	7.9	8.0	16.5	1.9
ATXW	9.8	10.0	10.0	17.5	2.6
ATYC	1.6	1.6	1.6	7.2	0.4
ATYD	4.8	4.6	4.8	13.6	1.8
ATYE	9.3	9.1	9.2	13.0	1.4
ATYF	2.2	2.3	2.2	12.7	1.5
AUQW	1.4	1.5	1.5	12.8	0.7
AUQW-B	0.3	0.3	0.3	11.5	1.6

CONCLUSIONS

Ion chromatography was proved to be an accurate and sensitive alternative method of analysis for inorganic anions and cations. The Dionex instrument is capable of multiple ion analysis from one aliquot. In addition, a wider range of concentrations can be determined by adjustment of attenuation of the conductivity cell. The Technicon analyzers are capable of analysis of 40 samples per hour with little operator attention. At present, one operator can analyze only 30 to 40 samples per day using two ion chromatography instruments. Therefore, the new method is not competitive at present for the large sample loads required by this laboratory. Future plans include automation of the instruments to increase the daily sample capacity.

ACKNOWLEDGMENTS

The authors wish to thank J. E. Bumgarner, R. L. Hedgecoke and C. M. Morris for aid in sample preparations and calculations.

REFERENCES

1. Mulik, J., R. Puckett, D. Williams and E. Sawicki. *Anal. Letters* 9(7):653-663 (1976).
2. Unpublished Technicon techniques used in this laboratory.
3. "Methods for Chemical Analysis of Water and Wastes," EPA-625/6-74003 (1974), p. 65.

7

APPLICATION OF ION CHROMATOGRAPHY
TO THE ANALYSIS OF ANIONS EXTRACTED
FROM AIRBORNE PARTICULATE MATTER

P. K. Mueller, B. V. Mendoza, J. C. Collins and E. S. Wilgus
Environmental Research and Technology
Western Technical Center
Westlake Village, California 91361

ABSTRACT

Samples of airborne particulate matter collected in the Los Angeles area have been analyzed for anions using liquid ion exchange chromatography (IC). Five anions have been identified. Information has been obtained on chromatograph parameters influencing trade-offs between resolution of peaks and speed of analysis. Various extraction procedures have been evaluated for extracting the ions from different sampling media. Several factors influencing the lower quantifiable limits of the detection method are discussed, such as water purity, room temperature requirements and sample size. This chapter also makes a cost comparison of routine workload analysis using the IC and automated colorimetric procedures.

IC is a very useful tool for the multi-ion analysis of extracts. Conditions for determining ionic substances at the highest detector sensitivity remain to be defined and established.

INTRODUCTION

The initial motivation for ERT to acquire ion chromatographic analysis capability was the need to determine aerosol sulfates on 10 to 30 minute collections of samples. The applications are in the analyses of filters

77

after aircraft sampling, diurnal sampling at rural sites and continuous sulfate analyzer testing.

In reviewing available methods, we considered the procedures listed in Table 7.1.[1-4]   Initially, we assumed that the flash volatilization method developed by Husar et al.[3] would be the most suitable. However, despite substantial trials we had difficulty in achieving the satisfactory precision and stability in calibration. Both the AIHL and Brosset methods were then left as potentially acceptable candidates together with IC. However, at the time when a selection had to be made, possible interferences with each method were under investigation, and it was reported that their accurate application required considerable experience. Also, their lower working limit is on the borderline of acceptability for our applications. In addition, IC has the capability of determining other ions from the same sample. We therefore adopted the IC for the determination of anions in our samples.

Table 7.1   Methods Used for the Quantitative Determination of Sulfate

Method	Lower Working Limit (μg/ml)
Methylthymol-blue[1]	6
Barium Chloranilate[1]	13
Barium Turbidimetric[2]	30
Brosset[1]	3
AIHL[2]	1
Flash Volatilization[3]	0.01
Ion Chromatography, Reported[4]	0.05
Ion Chromatography, ERT Experience	0.5

METHOD

A Model 10 Ion Chromatograph (Dionex Corp., Palo Alto, California) was used in the anion analysis of particulates collected on 47-mm TFE-coated glass fiber filters. The filter medium was obtained from Pallflex Products Corp., Putnam, Connecticut, under the product identification number EMFAB TX40A30. Subsequent batches of material with this ID No. showed particulate breakthroughs. We have now successful experience with a material designated TX40HI20 made to ERT specifications. Glassware used in the experiments was cleaned with detergent and rinsed ultrasonically with distilled deionized water. The filter (loaded or blank) was placed in a clean graduated 10-ml test tube; 5 ml of eluent buffer was added, and the tube was then capped. Anions were extracted ultrasonically for 30 minutes. The extract was filtered through a 0.2-μm membrane filter (Gelman Metricel) in a Swinny holder as it was injected into the chromatograph.

For qualitative determinations, the chromatograph was operated at the flow rate produced by 100 psi pressure at the pump. Standard solutions of suspect ions were injected individually; a mixture of standards was also injected. The anions were identified by matching the retention times of each peak in the unknown sample with each peak in the chromatogram of the standard mixture.

For determining only sulfate ($SO_4^=$) in the aqueous filter extracts, the instrument pump was adjusted to obtain pressures of 400 to 500 psi. The calibration standard was weighed and anhydrous Na_2SO_4 was quantitatively transferred for volumetric dilution with eluent buffer. Peak heights were measured from the linear baseline drawn from the inflection point of emergence of the first peak to that where the last peak terminates. Calibrations were based on peak heights obtained with standard $SO_4^=$ concentrations in μg/ml for each instrument range used. A $SO_4^=$ calibration was performed daily with each batch of samples. The $SO_4^=$ concentrations in sample extracts were determined from plots of μmho against $SO_4^=$ standards for each instrument range that was applicable.

The IC has been used to analyze 10- and 30-minute samples to complement data obtained with a laboratory prototype continuous $SO_4^=$ analyzer. The analyzer was being tested for one month in Downey, California.

RESULTS AND DISCUSSION

It was observed that temperature is a very important factor in performing IC analyses at the high sensitivity levels. Figure 7.1 shows the baseline drift as a function of temperature and electrical noise at an instrument setting of 1 μmho FS. Temperature fluctuations during an IC run would cause difficulty in peak identification and in assessing baseline, hence in peak measurements. When we installed the instrument in a temperature-controlled room ($\pm$ 2°C sudden variation), we were able to operate at 10 μmho FS without much trouble and at 3 μmho with more difficulty.

Negative peaks have been observed in chromatograms of samples extracted with distilled deionized water. This is shown in Figures 7.2, b and c. The negative peaks appeared where, or approximately where, the F^- and Cl^- peaks would emerge. This would have been a large source of error in the determination of these ions. Extraction with eluent buffer resulted in the elimination of the dips (Figure 7.2 a). The exact cause of the negative peaks has not been determined.

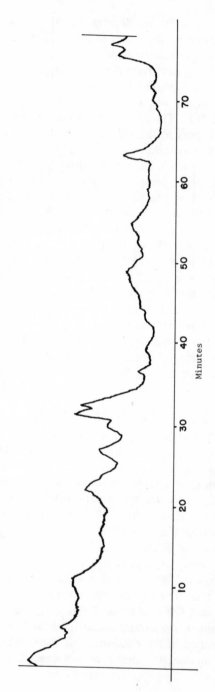

Figure 7.1   Zero drift at 1 μmho FS showing instrument noise and effects of about 2°C temperature fluctuations.

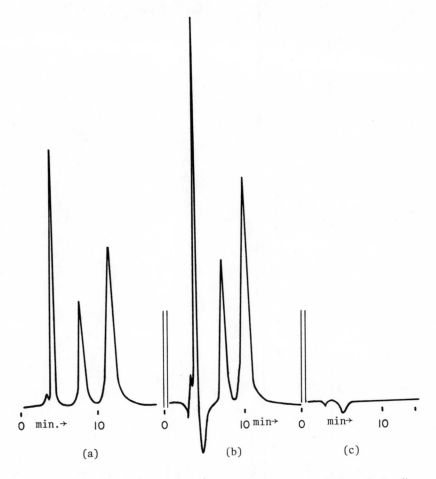

Figure 7.2   Ion chromatograms showing the influence of extractions on the baseline.

  a. Eluent (1.9 mM NA$_2$CO$_3$, 2.4 mM NAHCO$_3$) extract of a filter loaded with particulates from air in Thousand Oaks, California, at 30 μmho FS;

  b. Water extract of a filter loaded with particulates from air in Thousand Oaks, California, at 30 μmho FS; and

  c. Deionized water at 100 μmho FS.

A qualitative IC is shown in Figure 7.3 for the chromatogram of an extract of particulates collected on filters in Thousand Oaks, California. Seven peaks are shown. The peak at 8-minutes elution time coincides with the NO$_2^-$, but it is too weak to be certain. The peak at 22-minutes elution time was not identified. The first four peaks are characteristic of

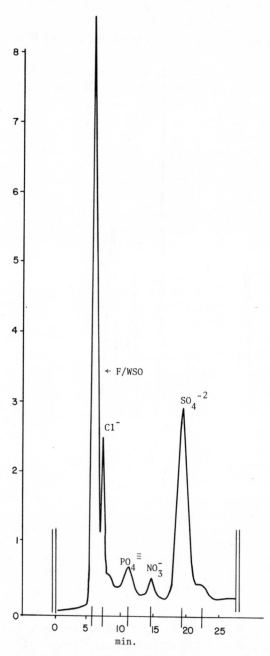

Figure 7.3   Anions extracted with eluent buffer (1.9 mM NA$_2$CO$_3$, 2.4 mM NaHCO$_3$) from particulates collected on filters in Thousand Oaks, California, on August 1, 1976. IC run at 10 μmho FS and 100 psi.

our filter medium, but very small quantities of Cl^- and F^- may also be present.

A quantitative evaluation of the identified anions is shown in Table 7.2 for samples collected during two consecutive days at selected times in the afternoon at Downey, California. Nitrates tended to be high in the late afternoon, while $SO_4^=$ was high early in the afternoon. Both clean and loaded filter extracts showed peaks where F^- or water-soluble organics (WSO), Cl^- and NO_2^- are expected. However, in this run, significant changes in the $SO_4^=$ and NO_3^- concentrations within a four-hour period are evident, showing the capability of our sampling and analysis approach to obtain anion data in a relatively short sampling time. The significance of the diurnal variation shown here cannot be interpreted without more extensive sampling. Other diurnal data for $SO_4^=$ and NO_3^- have been published previously (Hidy et al.[5]).

Table 7.2   Anion Concentration in $\mu g/m^3$ Found in Downey, California, on Relatively High Pollution Days. Samples were Taken on Teflon-Coated Glass Fiber Filters at 28.3 1/min for 30 Minutes

Filter Number	Starting Time (hr)	Peak 1 (WSO/F$^-$)[a]	Peak 2 Cl^-	Peak 3 NO_2^-	Peak 4 PO_4^{-3}	Peak 5 NO_3^-	Peak 6 $SO_4^=$
\multicolumn{8}{c}{IC Run of November 9, 1976}							
1	1255	4.94	3.27	3.03	< 1	6.37	11.2
2	1330	5.18	2.98	2.50	< 1	8.45	11.3
3	1425	5.19	1.96	1.96	< 1	8.69	11.75
4	1500	5.14	2.2	3.03	< 1	8.69	14.05
\multicolumn{8}{c}{IC Run of November 10, 1976}							
1	Scrubbed Air	4.58	2.08	1.43	< 1	1.19	1
2	Scrubbed Air	6.01	2.38	2.20	< 1	1.43	1
3	1325	5.24	1.90	1.96	< 1	10.77	22.5
4	1355	4.88	1.67	2.20	< 1	13.33	22.6
5	1430	4.46	1.73	2.50	< 1	13.09	18.1
6	1501	5.12	2.08	1.96	< 1	9.69	15.9
7	1533	3.99	2.14	1.43	< 1	15.17	14.8
8	1605	5.18	2.56	3.04	< 1	15.17	14.9
9	1637	5.24	2.50	3.33	< 1	14.64	14.4
10	Eluent Buffer Only	0.88	≤ 1	≤ 1	≤ 1	≤1	≤1

[a]$\mu g/m^3$ based on fluoride standard.

Data obtained simultaneously by the continuous $SO_4^=$ analyzer are plotted in Figure 7.4. Note the high correlation of the diurnal pattern.

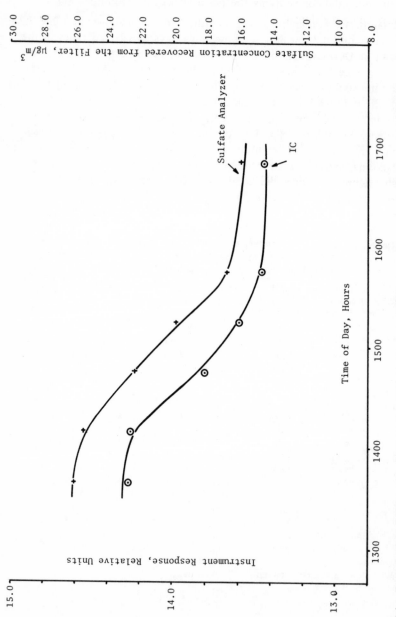

Figure 7.4   Diurnal distributions of sulfates as determined simultaneously by ion chromatography of 30-minute filter extracts and with a continuous sulfate analyzer on November 10, 1976.

The results show that IC coupled with short-term sampling on filters provides the potential for verifying the performance of continuous $SO_4^=$ analyzers being developed.

SUMMARY AND CONCLUSIONS

For achieving anion analysis on aqueous extracts of samples taken from aircraft, samples collected for short (30 minutes or less) time periods, and samples collected diurnally at rural sites, ion chromatography seemed to fulfill its expectations down to about 0.5 μg/ml. The IC accuracy with respect to standardized colorimetric methods remains to be established. Lower quantifiable limits are achievable with the reduction of filter background, solvent impurities and other sources of contamination, and with improved room temperature controls. However, these measures would also tend to make the method more tedious and costly.

REFERENCES

1. Appel, B. R., E. L. Kothny, E. M. Hoffer, G. C. Buell, S. M. Wall and J. J. Wesolowski. "A Comparative Study of Wet Chemical and Instrumental Methods for Sulfate Determination," presented before the Division of Environmental Chemistry, American Chemical Society, New Orleans, March 20-25, 1977.
2. Appel, B. R., E. L. Kothny, E. M. Hoffer, J. J. Wesolowski and R. D. Giauque. "A Comparative Study of Methods for Sulfate Analysis in Atmospheric Particles," *Proceedings of the International Symposium on Environmental Sensing and Assessment,* Las Vegas, Nevada, September 14-19, 1975.
3. Husar, R. B., J. D. Husar and P. K. Stubits. "Determination of Submicrogram Amounts of Atmospheric Particulate Sulfur," *Anal. Chem.* 47:2062 (1975).
4. Mulik, J., R. Puckett, D. Williams and E. Sawicki. "Ion Chromatographic Analysis of Sulfate and Nitrate in Ambient Aerosols," *Anal. Letters* 9:653 (1976).
5. Hidy, G. M. *et al.* "Characterization of Aerosols in California," Final report to the Air Resources Board of California, April 1975.

DISCUSSION

I noticed a negative peak in your chromatogram. Were you able to eliminate this disturbance?

We always found this peak in water extracts and were able to eliminate it when we used sodium bicarbonate buffer.

Is this negative peak commonly found in ion chromatograms?

Other workers have also observed them in their chromatograms. All water extracts give a negative peak.

Why did you extract your samples with carbonate buffer?

We extracted our samples with carbonate buffer to look at all the peaks. If we get negative peaks because of a water extraction, there is a chance that the peaks that coincide with the negative peaks will not come out, or will come out with shorter peak heights.

8

ION CHROMATOGRAPHIC DETERMINATION OF ANIONS COLLECTED ON FILTERS AT ALTITUDES BETWEEN 9.6 AND 13.7 KILOMETERS

D. A. Otterson

National Aeronautics and Space Administration
Lewis Research Center
Cleveland, Ohio 44135

ABSTRACT

The NASA Global Air Sampling Program (GASP) is designed to assess potential adverse effects of aircraft exhaust emissions to the lower stratosphere. Chloride, nitrate and sulfate, which are collected on cellulose fiber filters impregnated with dibutoxyethylphthalate (IPC 1478 filters) are among the pollutants being studied. An ion exchange chromatograph, which utilizes an anion exchange resin for the separation of these ions and a conductivity cell for their determination, provides a rapid and accurate method for the analysis of anions. Microgram quantities of the anions listed above are now being determined with an estimated sensitivity near 0.1 μg and a precision near $\pm$ 3 μg by NASA techniques. Evidence indicating the presence of other anions such as fluoride on these filters is discussed.

The sensitivity and precision were obtained only after procedures were developed to minimize contamination, to assure complete extraction of these anions from the filter, and to cope with side reactions that involve filter components. In so doing, the relationship between ion exchange phenomena and surface contamination of glass and cellulose became apparent.

INTRODUCTION

The NASA Global Air Sampling Program (GASP) is an investigation of atmospheric pollution at altitudes generally used by jet-liners. Part of this investigation is concerned with anion-containing particulates in the atmosphere at altitudes between 9.6 and 13.7 km. The samples are collected on cellulose fiber discs, which have been impregnated with dibutoxyethylphthalate. These filters have good retention for particulates and are designed for high-altitude air sampling.[1,2] Less than 30 μg of any one anion is collected on a sample because of limitations in the amount of air that can be filtered during a commercial jet-craft flight. Hence, very sensitive methods of analysis are required for this study.

Ion chromatography with eluent suppression and conductometric detection enables more thorough investigations of anions in the atmosphere than previous methods. This instrument was first described by Small *et al.*[3] They indicate that at least 29 anions can be detected in submicrogram amounts. Most of the anions can be determined in about 20 minutes in a single chromatogram.

A Dionex Model 10 Ion Chromatograph* designed for anion analysis is used in the GASP investigation. A similar unit and its use have been discussed in detail by Mulik *et al.*[4] Our preliminary work indicated that a sensitivity close to 1 ppb of an anion in solution should be obtainable. It is very difficult to attain these high sensitivities. Contamination and side reactions are the chief causes for loss of sensitivity. The factors that have been controlled to achieve the sensitivity needed for the determination of microgram amounts of F⁻, Cl⁻, NO_3^- and $SO_4^=$ will be discussed.

EQUIPMENT

Dionex Model 10 Ion Chromatograph

> Sample loop - 500 μl
> Separator column - 3 x 500 mm - filled with Chromex DA-x5-0.376 (127C) resin
> Suppressor column - 6 x 250 mm - filled with Chromex DC-x12-55 resin
> Pumping speed - 1.3 ml/min

Glassware

> 5-ml syringe - to transfer extract to sample inlet on ion chromatograph
> 10-ml syringe - to transfer eluent to extract ion bottle
> bottles - 30 ml - equipped with ground glass cap that fits around neck of bottle

---

*Mention of a specific product or company does not constitute endorsement by the National Aeronautics and Space Administration.

Hypodermic needle

3-in. - blunt end - with standard female Luer hub

Filter

The filter paper lint can be removed from the extract by a filter attached to the 5-ml syringe. One type is a wad of cotton placed on a wire coil inside the hub of the hypodermic needle. The other is a glass fiber filter without binder that can remove particles greater than 0.3 μm with 99.97% efficiency. It is held in a 13-mm diameter stainless steel syringe filter holder. It is less apt to become plugged by filter fibers than the first.

REAGENTS

Eluent - Mixture composed of 0.003 M NaHCO$_3$ + 0.0024 M Na$_2$CO$_3$
NaHCO$_3$ - Reagent grade
Na$_2$CO$_3$ - Reagent grade
Acetone - Electronic grade
Water - Deionized (resistivity > 1 megmho)

Organic materials removed by use of an activated charcoal column.
Air Stream - Purified in a column of molecular sieve (Linde 13 x) and a 10-μ mesh metallic filter

DECONTAMINATION OF EQUIPMENT

All surfaces that could possibly contaminate the extraction solution are cleansed and dried after each use. This includes the less obvious surfaces such as the ground glass surfaces of the bottle, its stopper and the syringe, as well as the hypodermic needle and the filter. These items are rinsed once with eluent, then three times with deionized water and finally with acetone. They are dried using a stream of purified air. Acetone should be removed as completely as possible. It can interfere with the Cl$^-$ determination. It produces a peak whose elution time is slightly greater than that of the chloride ion. This peak can overlap the Cl$^-$ peak if too much remains in the glassware. If the glassware contains relatively large amounts of anion, it may be necessary to use a stronger buffer to expedite its purification. In this even more extensive rinsing could be needed. During washing, the syringe plunger should be held by a metal holder to avoid contamination of the wash liquids by contact with one's fingers. Blanks are run each day to evaluate the cleanliness of the system. If a blank is too high, no determinations are made until blanks indicate that the source of contamination has been eliminated.

The filters (7.6-cm diameter) are purified individually in room air prior to being used to collect the sample from the atmosphere. Vacuum is used to remove the liquids from the filter resting on a sintered glass Buchner funnel. The purification steps include soaking in a carbonate buffer solution (0.3 M NaHCO$_3$ and 0.24 M Na$_2$CO$_3$) and then in 1 M acetic acid, rinsing thoroughly with six portions (30 to 35 ml) of deionized water which is saturated with dibutoxyethylphthalate. The filters are then dried at room temperature in vacuum over KOH pellets. Variations in cleanliness of the washed filters can result from differences in the contamination due to dust as well as differences in the treatment.

PROCEDURE

The analysis of the blanks and filters for anions is carried out in the same way except that one-fourth of a filter is added to the bottle before adding the eluent, if a sample is to be analyzed. The procedure is as follows:

Ten milliliters of eluent are added to the clean and dry bottle. The bottle is shaken, allowed to stand a few minutes and shaken again. Aliquots of 2.5 to 3.0 ml of the extract are drawn through the cotton filter in the needle into a clean and dry 5-ml glass syringe. This solution is used to rinse the inlet and sample loop of the chromatograph and to fill the loop. The contents of the 0.5-ml sample loop are injected into the eluent stream for the determination of anions. Strip chart recordings similar to the one in Figure 8.1 are obtained.

DISCUSSION

Investigations of the anion content of the atmosphere at altitudes between 9.6 and 13.7 km requires either more sensitive analytical techniques or much larger samples than are used for similar investigations of the atmosphere close to the earth's surface. Pollution of the upper atmosphere is much less than in the air at the lower altitudes. Ion chromatography is well suited for investigations of the anion content of the atmosphere at higher altitudes. Methods for the determination of a number of anions by this technique can be very sensitive and rapid contamination and interfering side reactions must be controlled to determine microgram quantities of anions collected on filters.

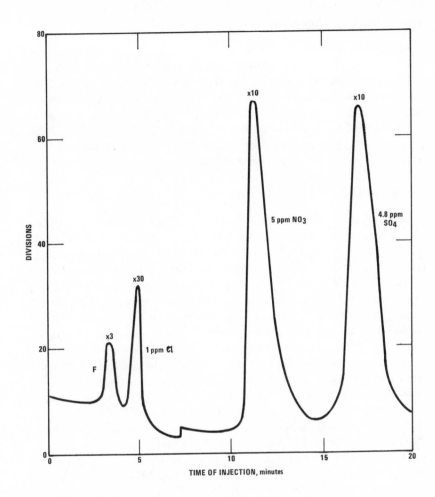

Figure 8.1 Typical ion chromatogram.

ION CHROMATOGRAPHIC DETERMINATION
OF ANIONS

The sensitivity and the speed with which anions can be determined in a solution by ion chromatography is indicated by Figure 8.1. It is a typical strip chart recording of the response of the ion chromatograph as a function of time. It reveals that F^-, Cl^-, NO_3^- and SO_4^{-2} can be determined in an aqueous sample in about 20 minutes. The height of the peaks implies that 0.001 ppm of an anion could be detected if the most sensitive setting (x .1) of the instrument were used.

Determination of Anions on Filters

The GASP method uses 10 ml of eluent to extract anions from one-fourth of the filter. The use of eluent for this purpose avoids interference with F^- + Cl^- that is caused by the water dip. This dip is obtained if the extracted sample contains considerably less carbonate and bicarbonate than the eluent. Inasmuch as the extract is injected into the ion chromatograph without further dilution, the detection limit of the method should be near 0.04 μg of an anion per filter. This sensitivity is difficult to attain because of contamination and interfering side reactions.

Contamination and Interfering Reactions

Most of the contamination and the interfering reactions occur during the extraction of anions from the filter and the transfer of the extract to the ion chromatograph. Extraction by the eluent appears to be complete in a few minutes. This extraction method was so effective that apparently clean glassware and filters were sources of contamination. The cleaning procedures, which were adequate for other determinations of anions on these filters, needed to be improved for this method. Experiments were carried out that led to a concept that explains a number of seemingly unrelated phenomena as well as to new purification methods.

The concept implies that exchange reactions that take place on anion exchange resins also occur on glass and on cellulose surfaces. All three surfaces contain more or less active hydroxyl groups (or other anions which had displaced these groups). These are assumed to be the active exchange sites. This similarity in surface compositions of these materials provides some theoretical support for this concept.

Two types of evidence suggest that exchange reactions occur on glass. In one instance, 10 ml of eluent removed 20 μg NO_3^- from the inner surface of a glass bottle after seven thorough rinsings with deionized water. Rough calculations indicate that all of this NO_3^- could have been adsorbed at hydroxyl exchange sites on the inner surface of the bottle. This suggests that competitive exchange between the carbonates and NO_3^- could have occurred at these sites and that the effective cleaning mechanism is due to carbonate ions displacing the NO_3^- from these sites. Similar behavior has often been observed after the bottle contained solutions with relatively high anion concentrations. That is, the eluent was able to remove measurable amounts of anions, which remained after thorough rinsing with water.

Table 8.1 presents data that could be due to Cl^- being retained on exchange sites. An HCl solution was added to platinum, glass and the filter. They were all vacuum-dried at room temperature. Both the filter

Table 8.1 Chloride Ion Retained by Platinum, Glass and IPC 1478 Filters
After Evaporation of HCl from their Surfaces

	Chloride Retained (μg)
Pt	~ 0.2
Glass	1.7
IPC 1478 Filter	1.5
Added as HCl	2.0

and glass retained measurable amounts of Cl⁻. Platinum, which has no surface hydroxyl groups, did not. These results suggest that a filter sample could also include a measurable amount of Cl⁻ from gaseous air contaminants. More experimentation is needed to establish the retention mechanism and to learn the capacity of the filters for the retention of this type of Cl⁻.

The similarity in behavior of glass and cellulose fiber filters such as indicated by this evaporation study is also found when the extraction of anions by deionized water is compared with that due to the eluent. As in the case of glass, extensive washing by pure water allows the filters to retain measurable amounts of anions. These are readily extracted by 10 ml of eluent. This supports the contention that competitive exchange reactions occur on filter surfaces. These reactions could readily account for the rapid extraction of anions from the filter by the eluent. Competitive exchange reactions on anion exchange resins are fundamental to ion chromatography. Ion chromatographic results may provide guidance in the treatment of glassware and cellulose. Indeed, the washing procedures for the glassware as well as the filters are based in part on this concept.

Table 8.2 indicates the degree of purity of the equipment that can be achieved by these cleaning procedures as well as the error that can be expected from differences. It presents the average values, $\bar{x}$, of the blanks for F⁻, Cl⁻, NO₃⁻ and SO₄⁻² obtained over a five-week period and the standard deviations for these values. Little error should be due to contamination of the equipment if the procedures are followed.

Table 8.3 reveals a much greater contamination of the washed filters as well as greater differences in purity of the various filters. It shows the average amount of F⁻, Cl⁻, NO₃⁻ and SO₄⁻² found on ten filters that were washed over a five-month period. Improved washing techniques are needed to reduce the error in analyses of the anions collected on filters.

Equipment to purify large numbers of filters simultaneously is now being constructed. Warm purified air is used to dry the filters. This is expected to provide uniformly pure filters in good supply. These will be

Table 8.2 Daily Blanks Obtained Between August 31, 1976, and October 4, 1976

	F^-	Cl^-	NO_3^-	SO_4^{-2}
		(μg per filter)		
$\bar{x}$	0.28	0.36	0.10	0.42
S	0.06	0.14	0.04	0.12

$$S = \sqrt{\frac{(\bar{x} - x_i)^2}{n - 1}}$$

where n is the total number of individual values, x_i.

Table 8.3 Anion Content of Ten Filters Purified Between August 28, 1976, and January 21, 1977

	F^-	Cl^-	NO_3^-	SO_4^{-2}
		(μg per filter)		
$\bar{x}$	1.28	0.88	0.34	1.13
S	0.66	0.98	0.24	0.36

needed for a more thorough study of the reliability of this method of analysis as well as for an improvement in the accuracy of analyses.

Interfering Reactions

Some organic compounds react with the eluent to form anions that can be detected by ion chromatography. Serious interferences with anion determinations may occur if organic compounds are present in the sample. For instance, low-molecular-weight alcohols can be detected if present in the eluent. Apparently, the alcoholate ions are formed. The interference due to acetone, which was mentioned previously, may be due to its enol form. The enolization of acetone can be written as

$$CH_3\overset{\overset{O}{\|}}{C}CH_3 + B \rightleftharpoons \left[CH_3\overset{\overset{O-}{|}}{C} = CH_2 \right]^- + BH^+ \tag{1}$$

where B is an alkaline substance.

Interferences with the F^- + Cl^- determinations are caused by reactions involving dibutoxyethylphthalate. These interferences were discovered when the peaks for these ions increased on aging of filter extracts. The NO_3^- and SO_4^{-2} peaks remained constant. It was soon learned that these interferences occurred with all solutions that contained both the eluent and dibutoxyethylphthalate. When a more dilute eluent was used, the elution times were extended sufficiently to indicate that the increase was caused by interfering substances. Dibutoxyethylphthalate is an ester. Esters saponify in alkaline solutions. If this ester saponifies in two steps, saponification products could be the sodium salt of butoxyethanol, sodium monobutoxyethylphthalate and sodium phthalate. The reactions that could yield these substances are

$$\text{(structure: ROOC, ROOC benzene)} + 2\,\text{NaOH} \rightleftharpoons \text{(structure: }^-\text{OOC, ROOC benzene)} + RO^- + H_2O + 2\,Na^+ \qquad (2)$$

$$\text{(structure: }^-\text{OOC, ROOC benzene)} + 2\,\text{NaOH} \rightleftharpoons \text{(structure: }^-\text{OOC, }^-\text{OOC benzene)} + RO^- + H_2O + 2\,Na^+ \qquad (3)$$

where R stands for the butoxyethyl group $(C_4H_9OC_2H_4)$. The sodium salt of butoxyethanol was found to elute with the F^- ion. Phthalate ion requires much longer elution times than any of the anions of interest. The elution time for monobutoxyethylphthalate has not been ascertained because none has been available. Some evidence indicates that the Cl^- interference decreases when phthalate ion increases. Hence, sodium mono-butoxyethylphthalate appears to be the cause of the Cl^- interference. If so, this implies that dibutoxyethylphthalate does saponify in two steps in a mildly alkaline solution such as the eluent.

Circumventing Saponification Interference

Because saponification is relatively slow, reasonably accurate results for $F^- + Cl^-$ can be obtained for IPC 1478 filters even though they do contain dibutoxyethylphthalate. Figure 8.2 shows the apparent increase in $F^- + Cl^-$ content as a function of time that elapsed between mixing of the eluent with a filter and injection of the extract into the ion chromatograph. These data indicate that errors due to saponification increase significantly with aging. The values indicated by extrapolation to the time of mixing is the amount of F^- or Cl^- on the filter if several assumptions are valid. The first assumption is that the rate of saponification is almost constant during the time period involved. The results obtained for three data points imply that this assumption is valid. The second is that no saponification products are on the filter at the start of the extraction.

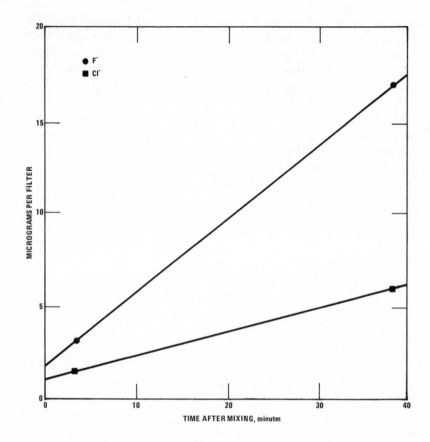

Figure 8.2   Extrapolation procedure for the determination of F⁻ and Cl⁻ in the presence of dibutoxyethylphthalate.

This means that the method cannot be used if IPC 1478 filters contain alkaline additives. For instance, tetrabutylammonium hydroxide has been added to some filters to improve their collection efficiency for acid gases such as HCl. Hence, these filters contain large amounts of the saponification products. The third assumption is that F⁻ and Cl⁻ are extracted quantitatively during the first aging period. One can infer that extraction of F⁻ and Cl⁻ is complete from the evidence that NO_3^- and SO_4^{-2} do not increase upon aging of the filter in the eluent. These two anions have much longer elution times than F⁻ and Cl⁻. If the anion exchange sites of cellulose and anion exchange resins are indeed the same, then these halide ions should be extracted from the filter faster than NO_3^- or SO_4^{-2}. This assumption is also supported by the data of Table 8.4. The

Table 8.4 Determination of F^- + Cl^- on IPC 1478 Filters

F (μg per filter)		Cl (μg per filter)	
Added	Found	Added	Found
10.2	9.2	10.7	10.5
10.2	9.4	10.7	10.5
		8.3[a]	8.3

[a]HCl added instead of NaCl.

agreement is close to the limits expected from the variation in anion content of washed filters (see Table 8.3). A more definitive study of the reliability of these procedures is being delayed until a large number of filters with uniform purity are available.

CONCLUSIONS

Ion chromatography is uniquely suited for the determination of anions collected on filters. It is rapid and thorough. An attribute of ion chromatography that is significant in environmental studies is its ability to indicate the presence of unsuspected anions. The presence of measurable amounts of F^- on our filters was not considered until it was identified as the cause of one of the peaks obtained in the analyses of the filter extracts. More work may be required to establish the validity of this identification. However, the possibility that F^- might be included in our investigation of the upper atmosphere may be quite significant in view of the Freon® controversy.

Contamination and interfering side reactions must be controlled to utilize the ultimate sensitivity of ion chromatography. The control of these factors to the extent necessary for the determination of microgram quantities of anions on cellulose filters has been discussed. Evidence was presented to support the view that glass and cellulose surfaces are sites of the same competitive anion exchange reactions that occur on anion exchange resins.

REFERENCES

1. Lazrus, A.L., B. Gandrud and R. D. Cadle. "Chemical Composition of Air Filtration Samples of the Stratospheric Sulfate Layer," J. Geophys. Res. 76(33):8083-8088 (1971).
2. Cadle, R. D. et al. "Relative Efficiencies of Filters and Impactors for Collecting Stratospheric Particulate Matter," J. Atmos. Sci. 30(5):745-747 (1973).

3. Small, H., S. Stevens and W. C. Bauman. "Novel Ion Exchange and Chromatographic Method Using Conductometric Detection," *Anal. Chem.* 47(11):1801-1809 (1975).
4. Mulik, J. *et al.* "Ion Chromatographic Analyses of Sulfate and Nitrate in Ambient Aerosols," *Anal. Letters* 9(7):653-663 (1976).

9

ANALYSIS OF ANIONS IN COMBUSTION PRODUCTS

R. D. Holm and S. A. Barksdale
Monsanto Research Corporation
Dayton Laboratory
Dayton, Ohio 45407

ABSTRACT

The objective of this work was to apply ion chromatography to measurements of acid gases present in the smoke from burning materials. The type of combustion products formed is strongly dependent upon the nature of the material and the burning conditions. Hydrogen halides may be formed in combustion products from PVC or certain flame retardant polymeric preparations. There was some concern over whether the heat of combustion products present would affect the anionic analyses desired. Interest centered on the analyses for Cl^-, PO_4^{-3}, Br^-, NO_3^- and SO_4^{-2} in smoke. Of these, Cl^- and Br^- were of greatest concern. The primary analytical problem in this work was the measurement of Cl^- and Br^- individually at low levels in the air. Ion chromatography provided an unusually convenient method for this purpose since most other procedures yield Cl^- levels by difference between Br^- and total halides.

Combustion experiments were conducted in an NBS Smoke Chamber. Samples burned included Douglas fir and a variety of polymeric materials. Acid gases were sampled using fritted bubblers located inside the smoke chamber, containing 25 ml of 0.003 M HCO_3^-/0.0025 M CO_3^{-2}, through which air was drawn at 140 ml/ min. Approximately 0.01 to 1 μmol of gas were collected, forming solutions of 10^{-6} to 10^{-4} M. A 1-ml sample loop was used. Measurements of Cl^-, Br^-, PO_4^{-3}, SO_4^{-2} and NO_3^- were feasible with this procedure. Sensitivity was such that 0.5-liter air samples containing 1 ppmv acid gas were measurable. Organic products have not produced any interferences. Large amounts of NO_x may produce sufficient NO_3^- to interfere with measurements of very small Br⁻ levels. This procedure was also used in studying the distribution of Br⁻ contained on soot deposited on the wall of the smoke chamber, and in other applications involving coal, leachate, carbon black and fly ash.

INTRODUCTION

Monsanto Research Corporation has been interested in fire performance and the chemical constitution of smoke for a number of years. For this purpose, we have been measuring certain gases continuously during combustion, including CO, CO_2, NO_x, O_2 and total hydrocarbons. Frequently, halogens are evolved during combustion, and the study to be described was directed toward the measurement of HCl and HBr in smoke products.

Our interest centered around polymers such as PVC, with flame retardant polymer additives. These materials release halogens as free radicals, which inhibit flame propagation. As a result we have been interested in measuring the evolution of HBr, HCl and sometimes HCN, during combustion. Ion chromatography provides an especially convenient method for the measurement of Cl⁻ in the presence of Br⁻ at concentrations of 10^{-6} to 10^{-4} M. Such measurements are difficult and imprecise using conventional colorimetric methodology. For example, the colorimetric Fe(III)-Hg(SCN)₂ method measures total halides, including CN⁻, if present. If Cl⁻ is desired, it must be obtained by difference after Br⁻ and CN⁻ have been measured independently and subtracted.

In applying ion chromatography to the measurement of halides, we were concerned about the effect that other smoke products might have upon the analyses. During the analysis of acid gases in combustion products, the following were observed:

1. Flaming Combustion - CO_2, NO_x, CO, SO_2, Halides
2. Nonflaming Combustion - (CO_2), CO, Halides, Organics

Smoke evolved during flaming combustion is usually high in CO_2, NO_x and SO_2 if sulfur is present. Halides are rapidly released. Nonflame combustion, for example smoldering, usually evolves much less CO_2 and NO_x, and considerably more CO and organic degradation products. Halides are again relatively easily evolved.

We have found that typical ionic products scrubbed from smoke include CO_3^{-2}, Cl⁻, PO_4^{-3}, Br⁻, NO_3^- and SO_4^{-2}. Our main concern initially was that the many organic products and NO_x present would interfere in the halide analyses we desired. In general, we have found that the organics do not interfere, and in most of our cases, the NO_x does not either.

EXPERIMENTAL DETAILS

The gas measurement apparatus used in our smoke studies is shown in Figure 9.1. Combustion experiments were carried out in an NBS Smoke Chamber, shown in Figure 9.2, containing 18 ft^3 volume, modified for continuous monitoring of several combustion gases. Polymer samples 3 sq. in. were exposed to a radiant flux of 2.5 to 5 W/cm^2 to simulate the radiant energy contribution of an adjacent fire. Samples were exposed on a horizontal burner assembly illustrated in Figure 9.3.

Figure 9.1 Continuous combustion gas measurement system.

Halides in the smoke were trapped by drawing gases through fritted bubblers containing 25 ml of scrubbing solution at 140 ml/min for 2 to 4 minutes. Figure 9.4 illustrates the interior of the chamber with scrubbers in place.

Figure 9.5 illustrates a flaming exposure with evolved smoke rising between the two heaters.

The scrubbing solution was the same composition as the chromatographic eluent used subsequently, 0.003 M HCO$_3^-$/0.0024 M CO$_3^{-2}$.

Figure 9.2 NBS smoke chamber.

Typically, 10^{-8} to 10^{-6} mol of sample was collected, producing 10^{-6} to 10^{-4} M concentrations. A 1-ml sample loop was used with the Dionex ion chromatograph since some concentrations were less than 0.1 ppm. The HCO_3^-/CO_3^{-2} eluent mentioned above was employed for most of the work reported here, except for some Cl^- analyses, for which a 0.005 M HCO_3^- eluent was used.

OBSERVATIONS

The three main observations I wish to speak of deal with (1) enhanced resolution at lower concentrations, (2) the effect upon a Br^- analysis of an increasing NO_3^- concentration and (3) an example of the ions measured in a typical experiment.

Figure 9.3   Horizontal burner assembly, equipped with flame igniters, installed in NBS smoke chamber.

Concentration Limitations on Resolution

We have observed, using a 1-ml sample loop, that resolution suffers greatly when ion concentrations greatly exceed 10^{-4} M. Figure 9.6 illustrates this effect for two concentrations of PO_4^{-3}, Br^- and NO_3^-. All three ions exhibit retention times in the same region. Two chromatograms, for a 10^{-3} and a 10^{-4} M concentration in each ion, are represented. It can be seen that the resolution of the 10^{-4} M ions is considerably better than that of the more concentrated solution. Resolution in chromatograms of the 10^{-5} and 10^{-6} M solutions was comparable to the 10^{-4} solution. When using an 0.1-ml sample loop, similar results were obtained, except that a 10^{-3} M solution in each ion could be resolved while a corresponding 10^{-2} M solution could not. We concluded that the capacity of the analytical column was several millimoles, and thus established the upper useful limit on concentrations of such neighboring and overlapping ions.

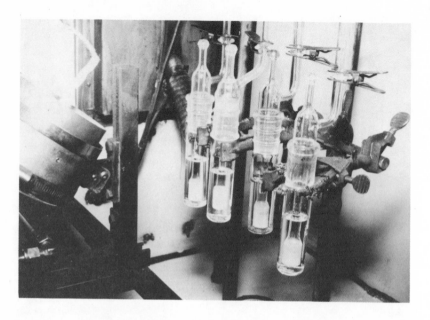

Figure 9.4   Acid gas fitted bubblers installed in center of NBS chamber for smoke sampling at different time periods.

BROMIDE MEASUREMENTS IN PRESENCE
OF NITRATE ION

Flaming experiments always involve some formation of NO_x, formed simply by high-temperature fixation of atmospheric nitrogen. Nitrogenous materials, when burned, yield considerable additional NO_x. Some of the NO_x appears as NO_3^- in the scrubber solutions and can pose an interference in Br^- measurements when the Br^- levels are small relative to NO_3^-.

We examined a series of solutions having a range of NO_3^-/Br^- mole ratio from 0 to 10 to ascertain whether areas or peak heights were better for making Br^- measurements, and at what mole ratio the NO_3^- caused a 10% error in the Br^- value. Figure 9.7 illustrates four chromatograms in which the NO_3^-/Br^- mole ratio is varied from 0.5 to 5, respectively. The Br^-, which precedes the NO_3^-, is 10^{-5} M (0.8 ppm) in all solutions. Peak heights were measured directly, and the areas used were those presented by the Hewlett-Packard integrator employed.

Figure 9.5 Typical smoke experiment under horizontal flaming exposure.

Figure 9.8 illustrates the area of the Br^- peak as the NO_3^- concentration was increased. The apparent Br^- peak area consistently increased at first as NO_3^- was added. At NO_3^- concentration twice that of Br^-, the area again equalled that obtained in the absence of NO_3^-. Higher NO_3^- concentrations caused the integrator to apportion progressively less area to the Br^- peak.

Peak heights of Br^- were also measured, and these are illustrated in Figure 9.9 in which the percent error in Br^- is plotted vs mole ratio of NO_3^-/Br^-. The Br^- peak height rose steadily as NO_3^- was increased, and at a NO_3^- mole ratio of 5, the Br^- peak height incurred a 23% positive error, while the Br^- peak area suffered a comparable 22% negative error. We have decided to use peak areas when possible for these ions.

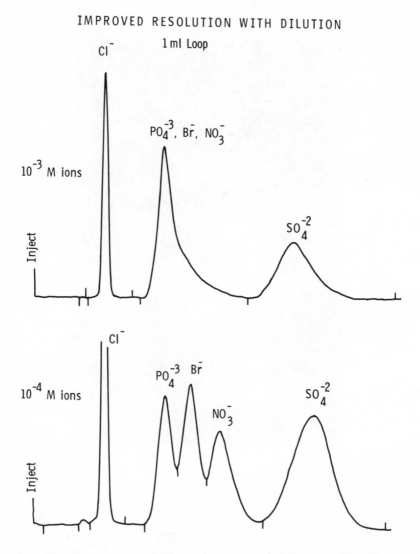

Figure 9.6 Illustration of observed resolution and, thus, maximum useful ionic concentration in a 1-ml sample loop.

A related situation arises when PO_4^{-3}-producing materials are used with Br^-. The PO_4^{-3} peak precedes the Br^- and poses an interference similar to that of NO_3^-. When this situation of poor resolution occurs, one can either switch to a smaller sample loop, or make a dilution.

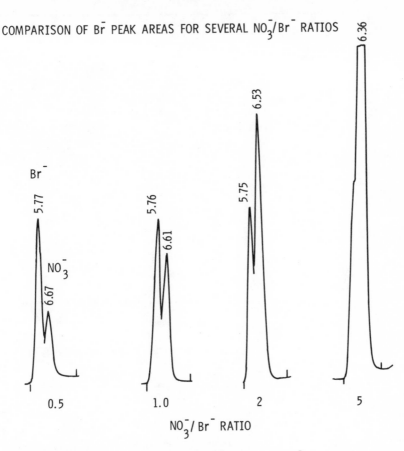

Figure 9.7 Comparison of ion chromatographic peaks for 10^{-5} M Br⁻ over a range of NO_3^-/Br^- concentration ratios.

APPLICATION TO POLYMER SMOKE CHARACTERIZATION

We have been interested in measuring halides in smoke. Figure 9.10 illustrates a typical chromatogram of the anions collected from the smoke obtained from a burning polymer containing Cl, Br and S. The sample was collected during the period from 6 to 8 minutes from the beginning of a nonflaming exposure; a considerable quantity of total hydrocarbons from pyrolyzed polymer was present. Such hydrocarbon concentrations can easily achieve 5000 ppm by volume in the air, but pose no problem in the analysis of Cl⁻ when it is measured using HCO_3^- eluent.

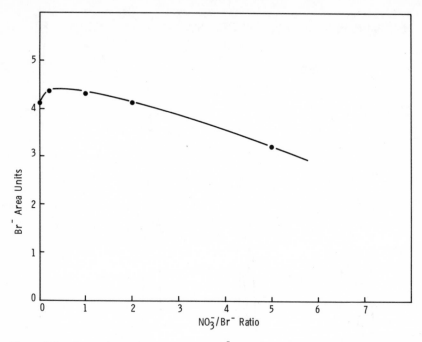

Figure 9.8   Comparison of peak area of 10^{-5} M Br$^-$ over a range of NO$_3^-$/Br$^-$ concentration ratios.

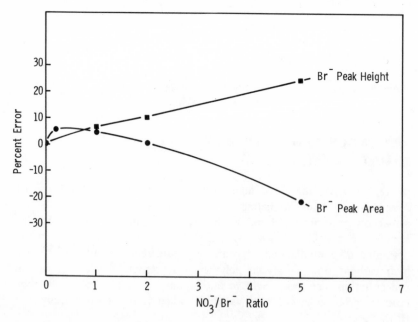

Figure 9.9   Percent error observed in Br$^-$ analyses using peak height and peak area measurements as a result of NO$_3^-$ overlap.

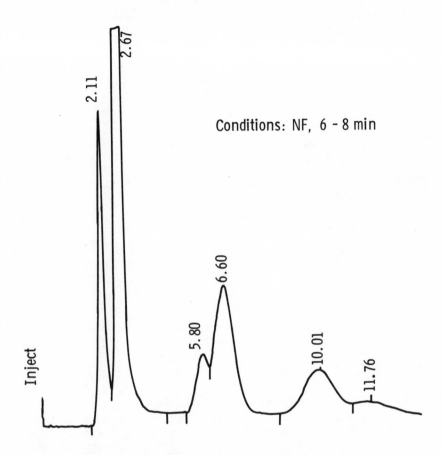

Figure 9.10 Chromatogram of anionic combustion products.

Hydrogen halides are very reactive and are actively scrubbed from the air by adsorption on soot, aerosol particles and chamber walls. Thus, the character of smoke can change significantly during the course of a 30-minute experiment. To demonstrate these effects, Figure 9.11 illustrates the change in observed Cl⁻ concentration in ppm by volume in smoke chamber air during the course of a flaming experiment. The airborne halide concentration passed through a maximum within a few minutes as deposition and abstraction rates began to exceed production rates. Similar behavior was observed for Br⁻.

The greatest benefit afforded by the ion chromatographic technique in our work is the ability to measure Cl⁻ directly, which was not previously available to us, and to provide rapid and sensitive measures for Br.⁻

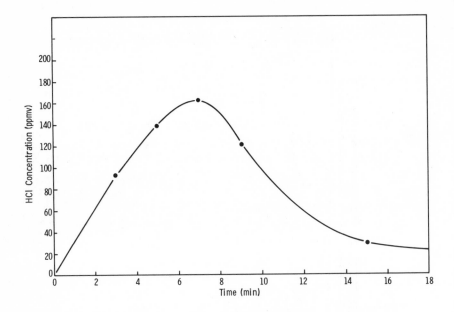

Figure 9.11 Typical variation in concentration of HCl in smoke during a nonflame combustion at 2.5 W/cm^2 radiant flux.

DISCUSSION

Was the burner assembly you used of the NBS type?

No. The horizontal burner assembly I illustrated was one we designed and fabricated. The burner assembly that comes standard with the NBS smoke chamber is a vertical assembly. We wanted to measure and look at the polymers that would run and drip and melt, and vertical burners are not designed for that.

In one of your chromatograms of combustion products, I noticed a small peak out beyond sulfate. Did you find out what that was?

No, we didn't. I have no idea what that was. We have seen a number of things that are very tightly retained. Iodide is an example. Our main interest was the halides.

Did you consider particulate chlorides on the filters?

We assumed that our samples were all particulate chlorides, and we found that the material dissolved very rapidly in the eluent.

10

MODIFICATION OF AN ION CHROMATOGRAPH FOR AUTOMATED ROUTINE ANALYSIS: APPLICATIONS TO MOBILE SOURCE EMISSIONS

S. B. Tejada, R. B. Zweidinger, J. E. Sigsby, Jr. and R. L. Bradow

Mobile Source Emissions Research Branch
Environmental Sciences Research Laboratory
U.S. Environmental Protection Agency
Research Triangle Park, North Carolina 27713

ABSTRACT

A Dionex Ion Chromatograph was successfully automated by interfacing to it an automatic sampler. Continuous unattended analysis of as many as 400 samples is possible, the actual number of samples being limited only by the ionic capacity of the suppressor column. The automated ion chromatograph (IC) was compared with the automated high-pressure liquid chromatographic (HPLC) adaptation of the barium chloranilate method in the analysis of soluble sulfates in stack and automobile exhaust samples. The two methods generally agree to within 1 μg, with filter samples containing less than 10 μg. Excellent agreement was obtained at higher concentration levels. Relative standard deviation was less than 5%. A modification, which would allow automated routine analysis of an unlimited number of samples, is also proposed.

INTRODUCTION

Four years ago, the measurement of sulfuric acid emitted by cars equipped with catalytic converters became a focal issue within EPA and the automotive industry because of the concern that the emitted sulfuric acid may potentially become a health hazard, especially in localities with high vehicular traffic densities. Analytical techniques available at that time lacked the sensitivity and precision for measurements of sulfates in solution at sub and low ppm levels. The urgent need for an accurate sulfuric acid emissions data base for both the noncatalyst- and catalyst-equipped cars led to the rapid development of fast, sensitive and precise methodologies for sulfates.

An automated high-pressure liquid chromatographic (HPLC) adaptation[1] of the barium chloranilate method (BCA) of Bertolacini and Barney[2] was developed at our Environmental Sciences Research Laboratory (ESRL). The minimum detectable quantity (MDQ) of sulfate by this method is less than 5 ng. Precision better than 3% at 0.5 μg/ml and better than 2% between 1.0 and 20 μg/ml has been attained. The automated BCA instrument is potentially capable of analyzing 20 samples per hour and up to 400 samples continuously without operator intervention. Butler and Locke[3] at Ford Motor Company developed a barium titration procedure using thorin as indicator with photometric detection of the end point. The sensitivity (defined as twice the titration blank) of this method is about 2.8 μg/ml, and the MDQ is about 10 μg. Both methods have been validated in a series of round robins and both are now widely used by laboratories doing automotive sulfate measurements.

The advent of ion chromatography (IC) after the publication of the Small et al.[4] paper on the first workable ion chromatograph opened a new and novel approach to multi-ion analysis. Mulik's demonstration of the first Dionex Model 10 unit at ESRL and his subsequent application of the technique to the analysis of nitrates and sulfates in ambient air filter samples[5] convinced us of IC's potential in the determination of ionic species in automobile emissions. We ordered a Model 10 fitted as an anion system and, along with it, a set of cation analytical-suppressor columns. Experience with our automated BCA sulfate instrument convinced us of the need to automate the IC to gain maximum use of the unit with the least expenditure of our limited manpower. We automated the Model 10 initially as an anion system by interfacing to it the sampling unit of our automated BCA sulfate instrument. For cation analysis, we initially used a Waters Model 6000A HPLC pump as an analytical pump and a Milton Roy mini-pump to regenerate the cation suppressor column. Switching between the cation and the anion mode of operation simply involved breaking and making connections in the columns. Repeated switching in this manner was a nuisance and eventually resulted in leaks at the affected hydraulic connections. It was at this point that we decided to modify the hydraulic and pneumatic plumbing of the Model 10 to make it convenient to operate, either as an anion or a cation system. In the modified unit, both the anion and the cation systems use the same sample injection valve and the same analytical and regeneration pumps. Switching from one system to the other is accomplished by a mere flip of an air toggle switch. What we have done is upgrade our Model 10 to extend its capability for automated routine applications. The modified instrument was used to measure sulfates in both stack and automobile emissions, and ammonia[6] in automobile exhausts.

INSTRUMENT MODIFICATION

Figure 10.1 shows the flow schematic of our automated IC shown as an anion system in the analysis mode of operation. Three manual 3-way Teflon valves, A, B and C (obtained from BioLab International) were installed before the analytical and regeneration pumps. A is used to select either special or routine eluents (anion, cation or water); B, to select regenerants (anion or cation); C, in conjunction with the syringe, is used initially to remove trapped air in eluent lines as well as to prime the analytical pump. The bubble trap between A and C is for removing air bubbles or dissolved gas in the eluents, which can cause the pump to cavitate and can result in erratic and/or complete stoppage of eluent flow. In automated operation where the instrument is left unattended for extended periods, removal of air bubbles is absolutely essential as pump cavitation may result in irretrievable loss of a large number of analytical results. This could be disastrous if the amount of sample is limited. We have not experienced any eluent stoppage attributable to pump cavitation since we installed the bubble trap.

The original slider valve syringe sample injection of the Model 10 was replaced by an air-operated 7000-psi Valco sample injection valve. A Technicon sampler-proportioning pump combination is used to load the sample into an external sample loop (100- or 500-μl volume). Two solid-state (automatic Timing and Controls Company) timers programmed for cyclic operation control both the sampler and the sample injection valve. In addition, the timers control the initiation of peak integration and printout of the calculated areas and/or concentrations at the end of each sample chromatographic run. The Technicon sampler used in the modified IC can hold 40 sample cuvets. Continuous analysis of up to 400 samples is possible by replacing the Technicon sampler with a Gilson sampler currently used in our automated BCA instrument. The actual number of analyzable samples is currently limited by the ionic capacity of the suppressor columns.

The original sample injection slider was used for plumbing the analytical columns, as shown in Figure 10.1. The slider valve used for flushing the pressure gauge was removed and combined with the original analytical

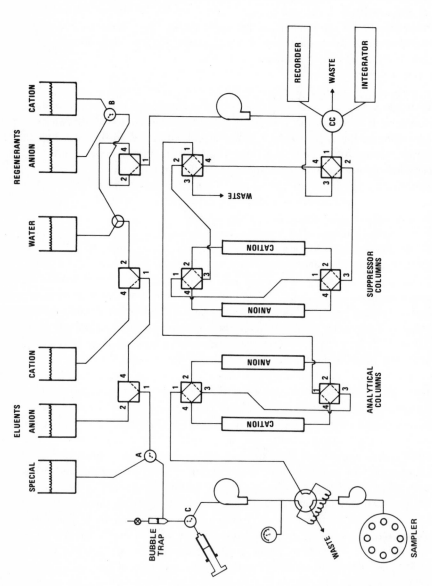

Figure 10.1   Schematic of automated ion chromatograph.

column slider valve. The suppressor columns were connected to this slider valve combination as shown in the figure. Actuation of the analytical and suppressor slider valves is controlled by a single air toggle switch. Pneumatic connections are such that only one analytical-suppressor column combination is possible for any one position of the air toggle switch, that is, the analytical-suppressor column pair must be either an anion or a cation column pair.

Selection of anion and cation eluents is still controlled by the original air toggle switches. With reference to the original Model 10 configuration, eluent E_2 is anion and E_1 cation in our modified IC. No modification was made on the slider valve-air toggle switch combination used for switching the suppressor column in series with the regeneration pump during regeneration and with the analytical pump during analysis mode of operation.

INSTRUMENT OPERATION

The operation of the instrument is similar to that of the Model 10 except that sample loading is done automatically.

Samples in suitable liquid medium are placed in plastic cuvets. The cuvets are then covered with polyethylene film to prevent solvent evaporation. In the load position the sampler needle probe pierces the polyethylene cuvet cover, and the probe dips into the sample so that its tip is about 1 mm above the bottom of the cuvet. Sample is pulled by peristaltic pump suction and pushed through the loop until at least two loop volumes of sample have passed through the loop. The sampling valve then switches to the inject position and the needle probe retracts to a deionized water reservoir. Water is continuously pumped through the sampling lines while the sample is being chromatographed to rinse off the residual sample. Switching of the sampling valve to the inject position initiates area integration of chromatographic peaks. At the end of the chromatographic run, area and/or concentration report is printed out, the valve switches to the load position, and the sampling process is repeated for the next sample.

EXPERIMENTAL

Stock solution of $SO_4^=$ (100.0 $\mu g/ml$) was prepared by dissolving 275.0 ± 0.1 mg of ultrapure $(NH_4)_2SO_4$ (obtained from Alpha Products, Danvers, Massachusetts) in doubly deionized water in a 2000-ml volumetric flask. The stock solution was immediately transferred into capped polypropylene bottles. Aliquots of this stock solution, containing 10 to

200 μg $SO_4^=$, were repeatedly dispensed into individual polypropylene bottles, and the solvent was evaporated in an oven at 80°C. These solid standards were dissolved in appropriate volumes of solvent to generate the sulfate calibration solutions for each series of analytical runs.

Most automotive sulfate samples are collected using the dilution tunnel technique. Basically, automotive exhaust is vented into a dilution tunnel where it is mixed with cool filtered air. SO_3 from catalytic conversion of SO_2 reacts rapidly with water in the exhaust to form sulfuric acid aerosols. The aerosols are collected on preweighed nominal 1-μ pore size Fluoropore filters downstream of the tunnel via isokinetic probes mounted in the flowing aerosol stream. Nominal sampling rate is 28 liters/min. The filters are equilibrated for 24 hours at 25°C and 40% relative humidity and then weighed to get the particulate mass load. The filters are then exposed to ammonia vapor for one hour to convert the sulfuric acid to ammonium sulfate, placed in polypropylene bottles and extracted with predetermined volumes of a 60% isopropyl alcohol water mixture for 20 seconds using a vortex test tube mixer. This extraction procedure was developed for the automated BCA method. The extract can be analyzed directly by this method. Sample preparation prior to IC analysis is outlined in Figures 10.2 and 10.3. A review of sulfate sampling methodologies and results of sulfate emissions measurements from catalyst-equipped cars was recently described by Bradow and Moran.[7]

RESULTS AND DISCUSSION

The performance of the automated IC was evaluated by repetitive analysis of aqueous sulfate solutions with concentrations of 0.5 to 20.0 μg/ml. Average relative standard deviation (RSD) of peak areas at six concentration levels was 2.2%. Scatter at 0.5 μg/ml was 4.5%. Figure 10.4 shows a typical chromatogram of standard ammonium sulfate solutions in water obtained with the automated IC. As can be noted in the figure, peak height reproducibility is excellent. Negative peaks preceding the sulfate peaks are due to deionized water used as sample solvent. The two spikes between successive sulfate peaks correspond to load and inject actuations of the sampling valve. The distance between the spikes is the duration of sample loop loading. We normally use a two-channel recorder to monitor the output of the detector, as shown in the dual chromatographic traces in Figure 10.4. The first channel is spanned to the detector output. In this manner, a more sensitive recorder response is obtained in the lower end of the sample concentration range. Precision and/or accuracy of peak height measurement at the low levels is thereby improved without the necessity of changing loop size or

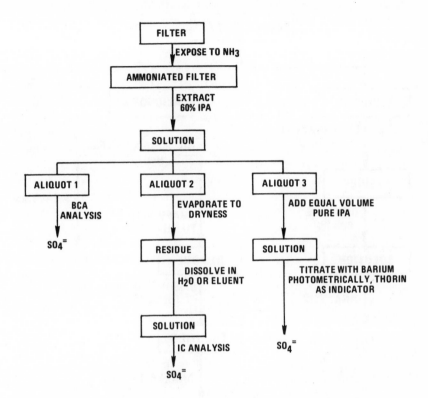

Figure 10.2 Analysis scheme for soluble sulfates in automobile exhaust.

detector sensitivity range. Figure 10.5 shows a concentration area calibration curve obtained with the automated IC. The regression line has an intercept of 0.0239, slope of 3.2978 and correlation coefficient of 0.9999.

In automated analysis involving a large number of samples, time lag between the time the samples are placed into the cuvets and the time they are analyzed may involve several hours. Solvent evaporation, if not controlled during this period, can result in significant error in observed concentrations of samples. To determine the effect of solvent evaporation on concentration, we ran a series of duplicate analyses of standard sulfate solutions at four concentration levels–0.5, 1.0, 5.0 and 10.0 $\mu g/ml$ during a period of eight hours. Four samples were analyzed at each of the concentration levels with the first two samples placed in plastic cuvets covered with polyethylene film and the next two in open plastic cuvets. The samples were arranged in the sample tray according to the pattern–covered, covered, open, open–for each concentration level. The samples were

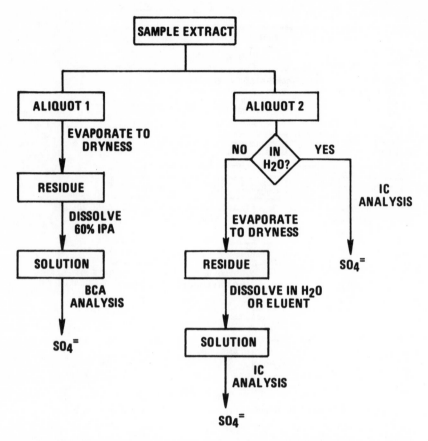

Figure 10.3   General analytical scheme for soluble sulfates.

analyzed sequentially and after the last sample was analyzed, the sample tray was turned back for a repeat analysis of the same samples. Time lag between successive samples was 15 minutes and between repeat analysis of the same sample, four hours. The results of this study are summarized in Table 10.1. The samples in the covered cuvets showed practically no change in concentration after four hours, while the samples in open cuvets showed as much as 15% increase in concentration. In the repeat analysis, the samples in the covered cuvets were partially exposed to the atmosphere as the plastic covers were already pierced by the needle probe during the first round. Evaporation loss under this condition appears to be insignificant. The results of this experiment emphasize the absolute necessity of minimizing solvent evaporation in automated analysis of multiple samples.

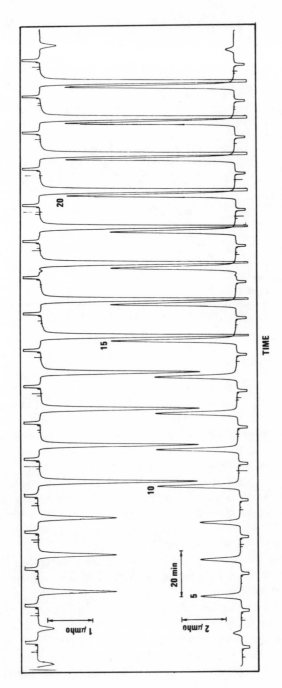

Figure 10.4 Chromatogram corresponding to repetitive injections of standard ammonium sulfate solutions in water.

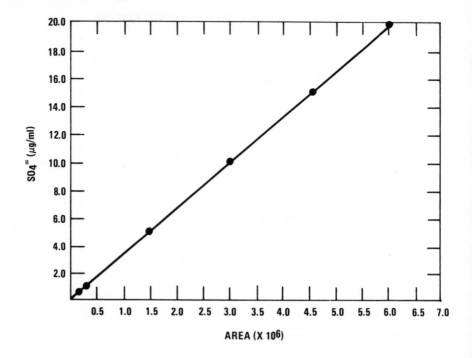

Figure 10.5   Peak area calibration curve.

Most of the sulfates in the diluted exhaust of catalyst-equipped cars are in the form of sulfuric acid and are routinely collected as such on Fluoropore filters. Because of the hydrophobic nature of the fluorocarbon surface, the extraction of sulfates from these filters necessitates the use of solvents, which wet the surface. Aqueous solutions of low-surface-tension alcohols, such as the 60% IPA used in the automated BCA procedure, are effective extraction solvents. However, caution must be exercised in direct injection of these alcoholic extracts into the IC columns as they may cause irreversible degradation of column performance. Such a case was encountered with 60% IPA.

The extraction procedure used in the comparative analysis of soluble sulfates in car exhausts by automated IC and automated BCA was the one developed for the latter, as shown in Figure 10.2. Conversion of sulfuric acid to ammonium sulfate by exposure to ammonia gas was observed to improve the precision in the automated BCA measurements, probably due to decreased adsorption of the sulfate species on the walls of the stainless steel tubing used in the BCA instrument. The filters were extracted with 10 ml of 60% IPA. Two 4-ml aliquots of each sample were pipetted into

Table 10.1   Effect of Solvent Evaporation on Sample Concentration

$SO_4^=$ Concentration (μg/ml)

Nominal	Observed		Cuvet	
	First Round	Second Round	No.	Condition[a]
0.5	0.50	0.54	1	X
	0.50	0.50	2	X
	0.53	0.67	3	O
	0.53	0.57	4	O
1.0	0.98	0.98	5	X
	0.99	0.97	6	X
	1.03	1.18	7	O
	1.02	1.18	8	O
5.0	4.80	4.71	9	X
	4.73	4.75	10	X
	5.10	5.72	11	O
	5.05	5.65	12	O
10.0	9.44	9.68	13	X
	9.54	9.55	14	X
	10.24	11.58	15	O
	10.53	11.62	16	O

[a]X = covered; O = open.

propylene bottles and the solvent was evaporated. One set of dried samples was extracted wtih 10 ml of the eluent or deionized water for IC analysis, and the other set was extracted with 10 ml of 60% IPA for BCA analysis. The original alcoholic extract could have been used directly for the BCA analysis as indicated in Figure 10.2.

Table 10.2 shows the results of the analysis of exhaust filter samples for sulfates by automated IC and BCA. Samples 6120 through 6129 were obtained from a car equipped with a three-way catalyst. The last four samples were laboratory-generated ammonium sulfate aerosol. The car exhaust samples show a low of 1.5 and a high of 10 μg of $SO_4^=$ per filter. The IC and the BCA results generally agree to within 1 μg per filter. Under the analysis conditions described in the previous paragraph, the actual concentration of sulfates in solution analyzed by either IC or BCA ranged from 0.07 to 0.4 μg/ml. The observed agreement is quite good considering the generally poor precision of peak area measurements at these low concentration levels. In the case of the ammonium sulfate aerosol samples, agreement between IC and BCA is excellent, the average relative percent difference being less than 2%.

Table 10.2 Comparative Analysis of Automobile Exhaust Filter Samples and
Generated Ammonium Sulfate Aerosol Samples by IC and BCA

	$\mu g\ SO_4^=$ per Filter		
Sample #	IC	BCA	Difference
6120	5.6	4.4	1.2
6121	8.7	10.0	⁻1.3
6122	8.4	7.9	0.5
6123	2.2	3.1	⁻0.9
6124	1.6	2.5	⁻0.9
6125	9.5	7.8	0.7
6126	6.7	7.1	⁻0.4
6127	4.3	3.4	0.9
6128	1.5	2.8	⁻1.3
6129	4.4	5.0	⁻0.6
RRw0B	164	156	8
RR21A	184	184	0
RR21B	160	158	2
RR93	93	94	-1

The stationary stack samples used in the IC-BCA comparison were in the form of water extracts of probe plugs, water rinses of the probes, 80% IPA absorbing solutions, and water extracts of the bubbler plugs. Sulfate concentrations determined by IC and BCA ranged from a low of 4.1 to a high of 321 $\mu g/ml$. The concentrations of most of these solutions were well above the sulfate concentrations usually observed in car exhaust extracts. Our automated IC and BCA instruments are normally set to analyze sulfate samples with an upper concentration limit of 20 $\mu g/ml$. High concentration samples were diluted as necessary. Sample handling prior to analysis by IC and BCA is detailed in Figure 10.3. Results of the analysis of the stack samples are given in Table 10.3. Agreement between IC and BCA is excellent, the average relative percent difference being less than 4%.

Sulfate is by far the predominant anion in practically all auto exhaust filter extracts we have analyzed so far by IC. Trace amounts of chloride and nitrate were sometimes observed. Chloride is scrubbed out in the automated BCA technique. Nitrate interferes positively in the BCA procedure but the interference is normally negligible at the concentration level nitrate is usually present in the exhaust filter extracts.

The results of our comparative analysis of auto exhaust filter samples and a few stationary stack samples lend credence to the equivalency of the automated IC and BCA procedures for soluble sulfates for these kinds

Table 10.3 Comparative Analysis of Stationary Source Stack Samples by IC and BCA

Sample #	Sulfate Concentration (μg/ml)		
	IC	BCA	Difference
24PP	60.2	59.7	0.4
24PW	16.9	16.2	0.7
241PA	52.7	54.5	-1.8
241P	4.1	4.2	-0.1
25PP	91.9	94.8	-2.9
25PW	10.9	11.6	-0.7
251PA	31.8	29.6	2.2
251P	4.2	4.3	-0.1
20877	321	309	11
21077	57.3	54.5	2.8

of samples. IC has distinct advantages over other methods for sulfates because it is essentially interference-free and more importantly has the capability to do multi-ion analysis. For relatively clean sulfate samples such as found in auto emission filters, automated BCA is a faster and more convenient method than automated IC. When a large number of samples needs to be analyzed solely for sulfates, automated BCA is usually employed in our laboratory. Only samples which show anionic interference are reanalyzed by IC. If only a few samples need to be analyzed, IC is normally used. This way, the IC is more available for use as a cation system. Application of IC to ammonia and amine analysis in auto exhaust is described by Zweidinger.[6]

One drawback of the present automated IC configuration for continuous routine analysis is that the number of analyzable samples is limited by the ionic capacity of the suppressor column. A simple extension of the modification scheme described in this chapter may remedy this deficiency. A pair of suppressor columns, both compatible with a given analytical column, can be installed in the unit at locations occupied by the anion-cation suppressor column pair in Figure 10.1. A timer programmer may be used to start the regeneration pump and to switch the suppressor columns after the injection of a predetermined number of samples. Instrument downtime will be minimized as this will involve only the time for baseline equilibration of the second suppressor column. In this manner, an unlimited number of samples can be analyzed continuously. This dual-suppressor column concept would be useful to laboratories doing routine analysis of a large number of samples of similar type.

REFERENCES

1. Tejada, S. B. "Determination of Soluble Sulfates in CVS Diluted Exhaust: An Automated Method," presented at NIEHS Symposium on Health Consequences of Environmental Controls, Durham, North Carolina, April 1974.
2. Bertolacini, R. J. and J. E. Barney, II. "Ultraviolet Spectrophotometric Determination of Sulfate, Chloride, and Fluoride with Chloranilic Acid," *Anal. Chem.* 30:202 (1958).
3. Butler, J. W. and D. N. Locke. "Photometric End-Point Detection of The Ba-Thorin Titration of Sulfates," *Environ. Sci. Health* A-11:79 (1976).
4. Small, H., T. S. Stevens and W. C. Bauman. "Novel Ion Exchange Chromatographic Method Using Conductimetric Detection," *Anal. Chem.* 47:1801 (1975).
5. Mulik, J., R. Puckett, D. Williams and E. Sawicki. "Ion Chromatographic Analysis of Sulfate and Nitrate in Ambient Aerosols," *Anal. Letters* 9:653 (1976).
6. Zweidinger, R. B., S. B. Tejada, J. E. Sigsby, Jr. and R. L. Bradow. "The Application of Ion Chromatography to the Analysis of Ammonia and Alkyl Amines in Automobile Exhaust," Chapter 11, this volume.
7. Bradow, R. L. and J. B. Moran. "Sulfate Emissions from Catalyst Cars, A Review," SAE Publication No. 750090, Detroit, Michigan (February 1975).

11

APPLICATION OF ION CHROMATOGRAPHY TO THE ANALYSIS OF AMMONIA AND ALKYL- AMINES IN AUTOMOBILE EXHAUST

R. B. Zweidinger, S. B. Tejada, J. E. Sigsby, Jr. and R. L. Bradow

Mobile Source Emissions Research Branch
Environmental Sciences Research Laboratory
U.S. Environmental Protection Agency
Research Triangle Park, North Carolina 27711

ABSTRACT

Ammonia present in automobile exhaust was trapped in dilute sulfuric acid and analyzed by ion chromatography. Alkylamines were also sought. A special low-capacity separator column was employed permitting a large number of analyses per regeneration cycle. The ammonia values are compared with those obtained by an on-line optical analyzer. Analysis for the methyl and ethyl amines is limited by sensitivity and interferences, and no amines were detected under the conditions employed.

INTRODUCTION

Statutory requirements for reduced carbon monoxide, hydrocarbon and oxides of nitrogen (NO_x) emissions from passenger cars has led to the development of emission control systems employing catalytic converters. Manufacturers have recently introduced the three-way catalyst system to reduce NO_x emissions. This system employs a platinum catalyst doped with rhodium to reduce NO_x to N_2. An oxygen sensor is used in the exhaust manifold to maintain a set air/fuel ratio through a feedback control circuit. The reduction of NO_x to NH_3 can occur if the air/fuel

125

ratio becomes too low, resulting in fuel-rich operation. This situation can occur if the oxygen sensor fails.

EPA has studied several prototype vehicles equipped with three-way control systems to assess any associated potential environmental risks. Normal and failure (oxygen sensor disconnected) modes of operation were investigated. Ammonia and other reduced species which could have been present were monitored. Ammonia emissions were studied by the two methods available—ion chromatography and second–derivative spectroscopy. The presence of alkylamines was also sought by ion chromatography.

EXPERIMENTAL

A Lear Siegler Model 1000 second-derivative spectrometer equipped with a multipass White cell was operated at 213.5 nm at a pathlength of 1 m. Samples were pulled through the instrument at 15 liters/min and continuously integrated with a Hewlett-Packard Model 3370A integrator. Output was displayed on a strip chart recorder. The sample stream was first passed through a 3 x 20-cm soda-lime scrubber to remove SO_2, which is an interference. The White cell, Teflon sampling lines and scrubber were all heated to 60°C. A commercial gas cylinder of 56 ppm NH_3 in N_2 was used for calibration.

A Dionex Model 10 ion chromatograph was equipped with a special 3 x 500-mm cation separator column and a 9 x 250-mm suppressor column. Samples were eluted with 0.0025 N HNO_3 at a flow rate of 2.45 ml/min. The instrument was modified for automated analysis as described by Tejada et al.[1] Samples were collected in bubblers containing either 15 or 25 ml of 0.01 N H_2SO_4 at a flow rate of 1.5 liters/min. Standards were prepared from stock $(NH_4)_2SO_4$ solutions.

Two 1977 California-standard Volvo automobiles equipped with three-way catalyst systems were used during this study. Testing was conducted in accord with Federal Register procedures[2] and included a particulate tunnel between the tail pipe and the constant volume sampler (CVS). Probes were extended into the particulate tunnel for sampling ammonia in diluted auto exhaust (Figure 11.1). The CVS was operated at approximately 400 ft^3/min. Two additional cycles, the Sulfate Emissions Test (S-7) and the Highway Fuel Economy Test (HWFET), were also used.

RESULTS AND DISCUSSION

The second-derivative spectrum of NH_3 is shown in Figure 11.2. The spectra of two known interferent gases present in auto exhaust, NO and SO_2, are also shown. Utilizing the 213.5 nm band for NH_3 measurements

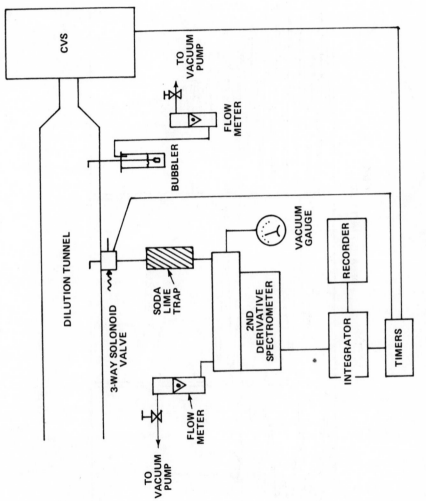

Figure 11.1 Auto exhaust sampling for ammonia.

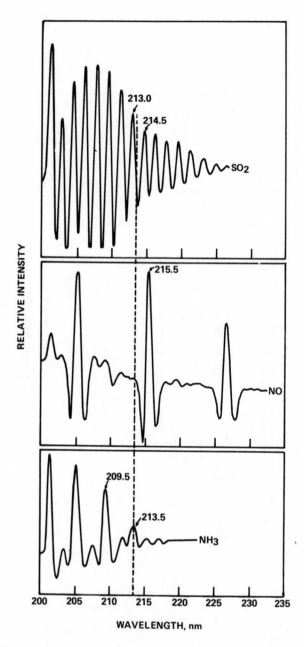

Figure 11.2   Second–derivative spectra of SO_2, NO and NH_3.

resulted in a 0.2% negative interference from NO. Since interference from SO_2 was about 150%, a soda-lime scrubber was placed in the sample line to remove SO_2. Other potential interferents were not investigated.

The determination of auto exhaust ammonia by ion chromatography necessitated trapping the ammonia as ammonium ion. Okita and Kanamori,[3] in studying atmospheric ammonia concentrations by the pyridine-pyrazolone method, reported ammonia collection efficiencies of 98 to 99% with bubblers containing 0.02 N H_2SO_4 at flows of 1 to 2 liters/min. Efficiencies of 97 to 98% were found for collection on glass fiber filters impregnated with 5 N H_2SO_4 at up to 15 liters/min. Since moderate levels of ammonia were expected and no sample preparation was required, bubblers containing 0.01 N H_2SO_4 were used in the present study. The 0.01N H_2SO_4 was also found to have collection efficiencies of 98 to 99%.

The determination of NH_4^+ by ion chromatography was initially carried out using the standard 9- x 250-mm cation separator column and 9- x 250-mm suppressor column with which our Dionex ion chromatograph was equipped. Elution with 0.01 N HCl gave good resolution with reasonable retention times. However, only about 10 samples could be run before regeneration of the suppressor column was necessary. Changes in response following regeneration also necessitated determination of a new standards curve. This resulted in only a few samples being analyzed each day. A special experimental cation separator column was subsequently obtained from Dionex. This was a 3- x 500-mm column in which the sulfonation level was approximately three times that of their normal cation resin. Satisfactory resolution and retention times were achieved using only 0.0025 N HNO_3 (Figure 11.3). Better resolution could probably be obtained using a smaller suppressor column.

The special cation separator column permitted a full eight hours of operation without regeneration. However, due to the column's relatively small volume and weak eluent requirements, sample pH had to be maintained at low levels. Figures 11.4 and 11.5 show the effect of increasing sample pH on peak geometry and area calibration, respectively. Samples were therefore collected in 0.01 N H_2SO_4. Low-level samples could then be directly injected with the 0.5-ml sample loop with minimal loss of resolution.

Figures 11.6 through 11.8 show the transient nature of ammonia in diluted auto exhaust for several different test procedures as measured by second-derivative spectroscopy. The low-range samples represent measurement made with the cars running properly, while the high-range samples represent failure mode operation (oxygen sensor disconnected). The Federal Test Procedure (FTP) is a three-part test consisting of a cold start phase (bag 1), stabilized running phase (bag 2) and, following a 10-minute

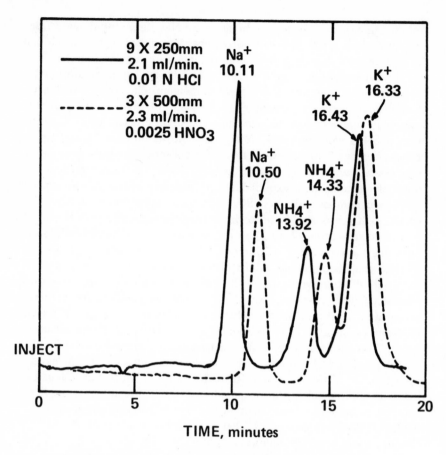

Figure 11.3   Comparison of cation exchange separator columns.

engine off period, a hot-start phase (bag 3). Overall values for the FTP are calculated by averaging 43% of bag 1, 100% of bag 2 and 57% of bag 3. Table 11.1, the ppm levels analyzed, shows fair agreement at high range, with greater scatter occurring at the low range. In both cases, however, values determined by ion chromatography were greater. Two sources of error were inherent in the second-derivative method under the conditions employed. Response time was greater than 30 seconds, which resulted in the integration of bag 1 being low and bag 2 being high. Also, Figure 11.6 indicates a negative signal response at the start of the FTP test. Although output is normalized to source energy, compounds present in unburned gasoline apparently absorb more UV energy than can be compensated for by the electronics in the second–derivative instrument. Any ammonia produced during this initial starting period would not be detected.

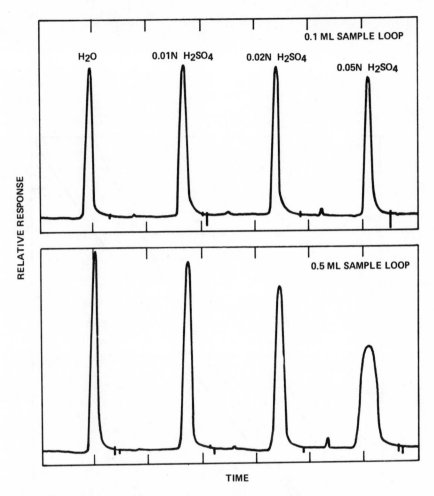

Figure 11.4 Effect of sample pH on peak geometry of NH_4^+ standards; 3- x 500-mm separator column with 0.0025 N HNO_3 eluent.

Two other tests conducted were the Sulfate Emissions Test (S-7) and the Highway Fuel Economy Test (HWFET). The S-7 test was developed to study sulfuric acid emissions from catalyst-equipped vehicles. It is a one-bag test procedure consisting of 13.5 miles of driving at an average speed of 35 mph. The HWFET simulates highway driving and is used to determine a car's "highway" mpg fuel consumption rating. It is also a one-bag test procedure and consists of 10.2 miles of driving at an average speed of approximately 50 mph. Both of these tests showed good agreement in the high range, but again exhibited poor correlation in the low range (Tables 11.2 and 11.3).

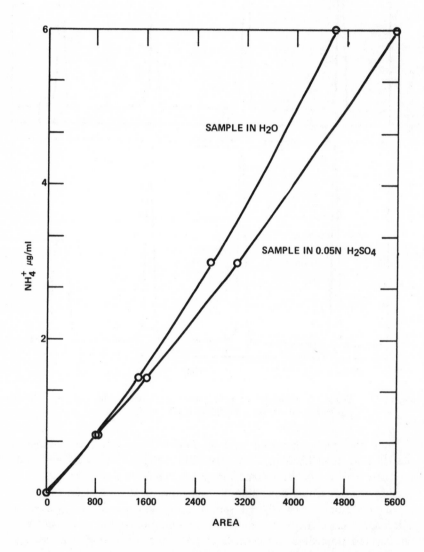

Figure 11.5 Area calibration curve for NH_4^+ using 0.5-ml sample loop; 3- x 500-mm separator column with 0.0025 N HNO_3 eluent.

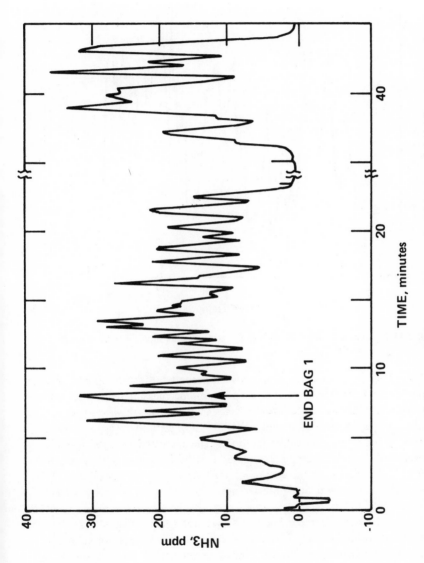

Figure 11.6 Ammonia emissions during FTP test cycle as monitored by second–derivative spectroscopy.

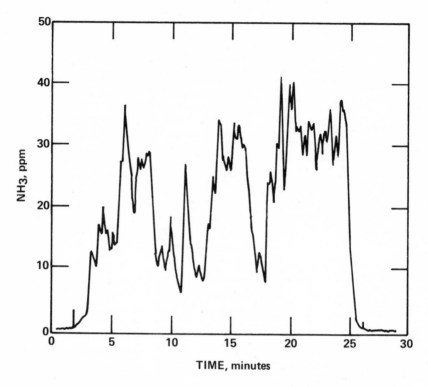

Figure 11.7 Ammonia emissions during S-7 test cycle as monitored by second-derivative spectroscopy.

The generally poor agreement between second-derivative and ion chromatography methods in making low-range ammonia measurements has several possible explanations. Interfering species may become significant at low levels. As previously stated, only NO and SO_2 were investigated as interferents in the second-derivative method. The increased ratio of second derivative to ion chromatography values for the low-range S-7 measurements (Table 11.2) relative to the low-range HWFET measurements (Table 11.3) may indicate a cycle-related interference. The necessity of using a soda-lime scrubber to pretreat the sample may also be a source of error. Factors such as variations in sample humidity have not been evaluated. A decreased signal-to-noise ratio at low levels may also result in integration errors.

Ammonia measurement by the ion chromatographic method appears a bit more straightforward but is not without sources of error. Methanol and ethanolamine are compounds known to elute with retention times similar to that of ammonium ion. The likelihood of these compounds being present in significant amounts is considered slight, however. Contaminated

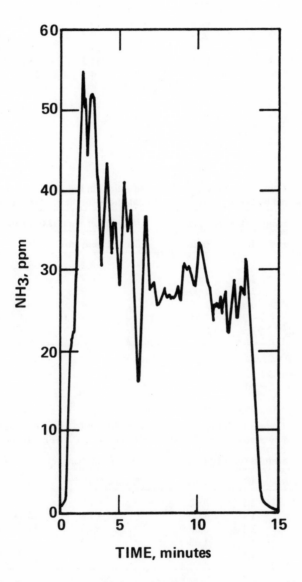

Figure 11.8 Ammonia emissions during HWFET test cycle as monitored by second-derivative spectroscopy.

Table 11.1 Comparison of Ammonia Concentrations in Diluted Auto Exhaust as Measured by Second-Derivative and IC Methods for the FTP Test Cycle

Low-Range Samples (0 to 10 ppm)[a]			High-Range Samples (0 to 100 ppm)[b]		
Second-Derivative	IC	Ratio	Second-Derivative	IC	Ratio
1.4	1.9	0.74	20.2	21.9	0.92
1.2	1.7	0.71	19.4	20.9	0.93
0.8	0.9	0.89	18.4	19.5	0.94
0.7	0.9	0.78	15.3	18.0	0.85
			14.5	16.9	0.86

[a]$\bar{R}$ = 0.78; RSD = 10.3%.
[b]$\bar{R}$ = 0.90; RSD = 4.8%.

Table 11.2 Comparison of Ammonia Concentrations in Diluted Auto Exhaust as Measured by Second-Derivative and IC Methods for the S-7 Test Cycle

Low-Range Samples (0 to 10 ppm)[a]			High-Range Samples (0 to 100 ppm)[b]		
Second-Derivative	IC	Ratio	Second-Derivative	IC	Ratio
1.8	1.0	1.8	38.2	38.5	0.99
2.2	2.1	1.05	30.9	30.4	1.02
2.0	2.0	1.00	34.2	32.2	1.06
1.1	0.8	1.38	18.4	18.3	1.00
2.0	1.3	1.54	26.5	27.5	0.96
1.4	1.5	0.93			

[a]$\bar{R}$ = 1.28; RSD = 27.0%.
[b]$\bar{R}$ = 1.01; RSD = 3.6%.

Table 11.3 Comparison of Ammonia Concentrations in Diluted Auto Exhaust as Measured by Second-Derivative and IC Methods for the HWFET Test Cycle

Low-Range Samples (0 to 10 ppm)[a]			High-Range Samples (0 to 100 ppm)[b]		
Second-Derivative	IC	Ratio	Second-Derivative	IC	Ratio
2.3	2.5	0.92	61.8	64.3	0.96
1.4	1.4	1.00	47.2	45.1	1.05
1.8	2.1	0.86	54.3	57.1	1.15
1.3	1.2	1.08	29.9	27.3	1.09
0.9	1.6	0.56	40.5	41.9	0.97

[a]$\bar{R}$ = 0.88; RSD = 22.5%
[b]$\bar{R}$ = 1.03; RSD = 7.7%.

glassware is always a potential source of error but can be minimized by careful cleaning and handling procedures. Impingers or bubblers constructed from polypropylene would be desirable for reducing background.

The presence of alkylamines was also sought by ion chromatography. The high concentrations of ammonia observed under certain test conditions made the presence of alkylamines probable. Alkylamines are only moderately toxic but may also be precursors for compounds like dimethylnitrosamine, a known carcinogen. Figure 11.9 shows the separation of the methyl- and ethylamines by ion chromatography. Dimethyl- and ethylamine are not resolved under the conditions employed. Also, potassium ion (Figure 11.3) is a potential interferent for methylamine. Figure 11.10 shows an actual exhaust sample run at high sensitivity, which contained approximately 50 μg/ml of NH_4^+. A 33.75 μg/ml NH_4^+ standard doped with 0.55 μg/ml of dimethylamine is also included for comparison. The exhaust sample showed no shoulder inflections that might indicate the presence of an alkylamine. The concentration of any alkylamines present must therefore be several orders of magnitude less than the ammonia.

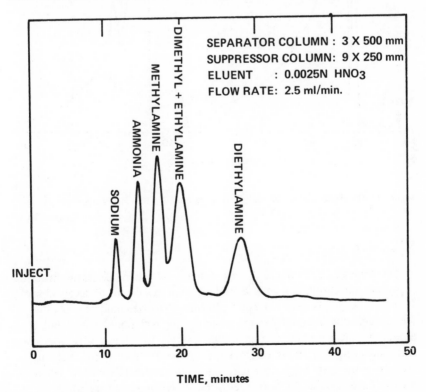

Figure 11.9 Separation of amines by ion chromatography.

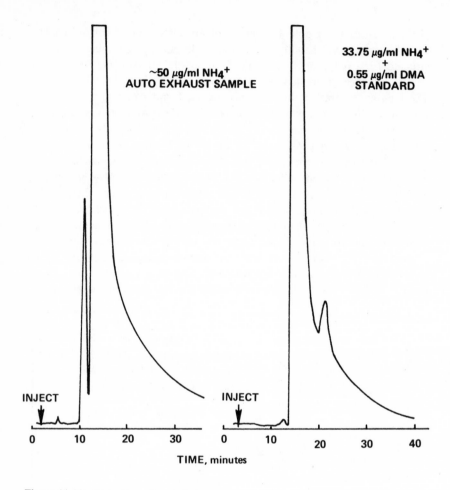

Figure 11.10   Detection of amines in auto exhaust by ion chromatography, compared to doped standard.

SUMMARY

The analysis of ammonia in auto exhaust by second-derivative spectroscopy and ion chromatography exhibited good agreement at levels above 10 ppm. At levels below 3 ppm, agreement was generally poor. The second-derivative method has the advantage of real time analysis but is limited by the necessity to pretreat the sample to remove SO_2 interference and, under the conditions employed, its response time is too long to accurately integrate sequential parts of a multiphase test cycle. Other interfering species might have been present but were not identified.

The ion chromatographic method for ammonia in automotive emissions appears free from interference and has a wide dynamic range. Routine analysis was best attained by selecting chromatographic conditions that allowed a high sample turnover per regeneration cycle. The ion chromatographic method for the analysis of alkylamines is hindered by poor resolution and interferences. Under the conditions employed, none were detected.

REFERENCES

1. Tejada, S. B., R. B. Zweidinger, J. E. Sigsby, Jr. and R. L. Bradow. In *Proceedings of the Symposium on Ion Chromatographic Analysis of Environmental Pollutants,* Environmental Protection Agency, Research Triangle Park, North Carolina (Ann Arbor, Michigan: Ann Arbor Science Publishers, Inc., 1977).
2. *Federal Register* 37(221):Part II (November 1972).
3. Okita, T. and S. Kanamori. "Determination of Trace Concentrations of Ammonia in the Atmosphere Using Pyridine-Pyrazolone Reagent," *Atmos. Environ.* 5:621 (1971).

12

APPLICATION OF ION CHROMATOGRAPHY TO STATIONARY SOURCE AND CONTROL DEVICE EVALUATION STUDIES

R. Steiber and R. M. Statnick

Industrial Environmental Research Laboratory
U.S. Environmental Protection Agency
Research Triangle Park, North Carolina 27711

ABSTRACT

The accurate measurement of sulfite and sulfate concentrations in flue gas desulfurization systems is extremely difficult. Using currently available analytical techniques, total sulfur as sulfate and total sulfite sulfur are measured directly. Sulfate sulfur is determined by difference. The ion chromatographic approach offers a technique for direct, simultaneous determination of sulfite and sulfate in process liquor streams. A comparison of ion chromatographic results and classical methods has been performed on samples obtained from a flue gas desulfurization process.

A second application which has been pursued is the direct determination of primary sulfuric acid emissions from coal-fired utility boilers. The sample is collected, utilizing the controlled condensation SO_3 sampling technique, and analyzed using ion chromatography. Comparison of the ion chromatographic results and the barium ion titration to the thorin end-point results has been excellent. Data from more than 50 separate measurements will be presented.

INTRODUCTION

The accurate measurement of the oxides of sulfur in both aqueous and gaseous streams presents special difficulties. In the case of scrubber liquor from flue gas desulfurization processes, the concentration of both sulfite and sulfate must be determined quickly and accurately. At present, this is accomplished by lengthy wet chemical techniques.[1] The time between acquisition of the samples and the actual completion of the analyses may be many hours. As a result, process operators are often denied the vital information they need to correct imbalances in the desulfurization system. There is also the problem of oxidation. The longer the sulfite sample remains unanalyzed, the more likely it is that a significant portion of it has oxidized to sulfate, thereby making accurate analysis impossible.

In the case of the measurement of sulfur oxides in stationary source gas streams, the problem is one of interferences. When a simple acid-base titration is used for the analysis, the presence of nitrates, chlorides, etc., will render the determination inaccurate. Even in the case of the barium ion titration, which is much more selective as regards sulfate, the presence of chloride, carbonate and sulfite ions in sufficient concentrations will cause positive errors in the results.

Recent studies[2] in our laboratory indicate that the application of ion exchange chromatography to these problems may result in their solution. In the case of sulfite and sulfate determinations, this methodology meets all the criteria for speed, accuracy, reproducibility and freedom from interferences. It may also be used for the simultaneous determination of chlorides, fluorides and nitrates in the sample.

SULFATE/SULFITE DETERMINATIONS
IN FGD* SYSTEMS

The instrument used in the following tests was a Dionex System 10 Ion Chromatograph.** It was equipped with a 500-mm anion analytical column followed by a cation suppressor column. The pump rate for all analyses was 1.5 ml/min and the sample size was 100 μl.

A study[3] conducted by SAMB-ESRL of EPA has already demonstrated the usefulness of ion chromatography in the determination of sulfate concentrations. It was determined in our laboratory that by manipulating the molarity of the eluent it was possible to resolve the sulfate and sulfite peaks. The manufacturer recommends that a solution containing 252 mg/l $NaHCO_3$ and 254.3 mg/l Na_2CO_3 be used as the eluent when a 500-mm anion analytical column is installed. By halving these concentrations, a complete separation of the sulfite and sulfate peaks was achieved. In

---

* FGD = Flue gas desulfurization.
** Mention of commercially available products is for identification only and does not imply their endorsement by the United States Environmental Protection Agency.

these tests, the retention time for sulfite was 23 minutes and the retention time for sulfate was 45 minutes (Figure 12.1).

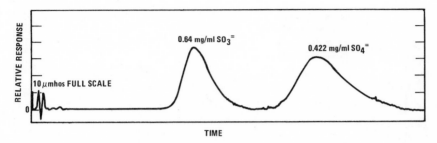

Figure 12.1 Separation of sulfate from sulfite.

Having obtained a suitable peak separation, the method was then tested for accuracy and reproducibility. A series of standard sulfite samples was prepared and injected onto the ion chromatographic column to generate a calibration curve. A separate sulfate curve was also generated. The range chosen for these samples is typical of sulfite and sulfate concentrations found in the scrubber liquor of the flue gas desulfurization system operated by TVA at their Shawnee Power Plant. In the case of both species, a linear, reproducible calibration curve was obtained. When standard concentrations of $SO_3^=$ and $SO_4^=$ were mixed together in varying proportions, there was no observable effect on either calibration curve. The curves are shown in Figure 12.2.

A series of nine separate analyses were made at each preselected sulfite ion concentration and sulfate ion concentration. Table 12.1 indicates the mean value and the standard deviation of the resultant integrated outputs. In addition to performing these separations on synthetic samples, a series of liquor samples from IERL/RTP's pilot-scale flue gas desulfurization system was run.

A portion of the liquor samples was injected onto the column and analyzed using the method of standard additions. Another portion of each sample was treated with hydrogen peroxide to oxidize all the sulfite to sulfate, acidified, and purged with air to remove carbonate ion. This oxidized portion was divided into two parts and analyzed by barium ion titration to the thorin end-point and by turbidimetric procedures. The total sulfate concentrations as determined by each method agree within the experimental error of the analyses. The results appear in Table 12.2.

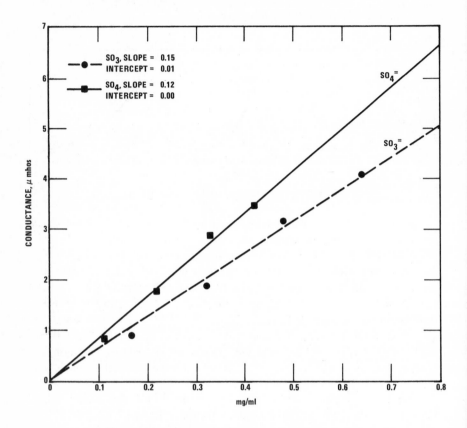

Figure 12.2 Calibration curves for sulfate and sulfite.

Table 12.1 Reproducibility of Ion Chromatographic Analyses of Sulfite and Sulfate

$SO_3^=$		$SO_4^=$	
Concentration (mg/ml)	Area	Concentration (mg/ml)	Area
0.64	4.43 ± 0.12	0.24	1.08 ± 0.00
1.6	13.30 ± 0.17	0.48	2.29 ± 0.09
2.24	20.00 ± 0.5	1.20	5.08 ± 0.05
3.20	30.5 ± 1.46	1.92	7.48 ± 0.05
4.80	440 ± 1.00	2.40	9.25 ± 0.05

Table 12.2 Analysis of FGD Liquor[a]

Ion Chromatography			Ba^{++} Titration	Turbidimetric[b]
SO$_3^=$	SO$_4^=$	Total SO$_X$ as SO$_4^=$	Total SO$_X$ as SO$_4^=$	Total SO$_X$ as SO$_4^=$
2.0	2.1	4.5	4.4	4.2
1.5	2.7	4.5	3.7	4.0
0.0	2.4	2.4	2.5	2.5
0.0	2.1	2.1	2.0	1.7
0.7	3.3	4.1	4.1	4.1
0.0	1.4	1.4	1.5	1.3
0.8	1.8	2.7	2.6	2.7
0.0	1.1	1.1	1.2	1.1

[a]Results are given in mg/ml.
[b]Adjusted for negative bias of method.[4]

DETERMINATION OF SULFURIC ACID MIST IN GAS STREAMS

The application of the ion chromatographic technique to the problem of sulfuric acid mist sampling and analysis has proved successful. In this context it has been used by our laboratory as both an analytical and a quality control system.

The preferred method[5] for the acquisition of SO$_3$ samples from stationary source gas streams is by the controlled condensation technique. To effect this condensation, the coil temperature must be maintained at 60°C. A significant variation from this temperature may result in the condensation of other flue gas constituents, which will then interfere with the analysis.

Quality control checks by ion chromatography on condensation coil washes taken at the Shawnee Power Plant were able to demonstrate that on-site personnel were carefully following procedures developed for them by our laboratory. The samples tested contained only sulfate and an insignificant amount of chloride (<0.01 ppm), the latter due to impurities in the power plant distilled water supply (Figure 12.3). This rapid, accurate quality control check was only possible because the ion chromatograph "sees" not only sulfate ions, but also chloride, fluoride, phosphate, nitrate, etc., as well.

In conjunction with the above quality control checks, a comparison of ion chromatograph results was made with the barium ion titration. In these tests, standard concentrations of H$_2$SO$_4$ were analyzed by both techniques. The barium ion titration results generally agreed with the

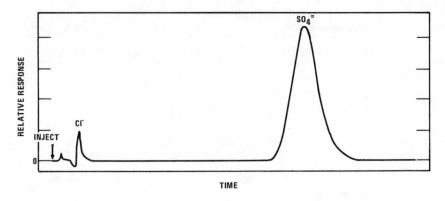

Figure 12.3   Goksoyr-Ross coil sample from Shawnee Power Plant.

known $SO_4^=$ concentration; the ion chromatograph results were well within experimental tolerances.   The data are shown in Table 12.3.

Table 12.3 Comparison of IC Method with Barium Ion Titration Using Standard H_2SO_4 Solutions Normality

Actual	Barium Ion	IC
0.005	0.005	0.006
0.010	0.010	0.011
0.025	0.025	0.026
0.040	0.040	0.040
0.050	0.051	0.049

CONCLUSION

This chapter has discussed only a narrow range of the many possible applications of ion chromatography to the analytical demands of stationary source evaluations.   For instance, current laboratory procedures for the determination of chloride and fluoride are long and involved and subject to a wide range of error by technical staff.   By the use of ion chromatographic techniques, however, both chloride and fluoride can be run simultaneously in a matter of minutes.   It may also be possible to develop a method for the direct, simultaneous analysis of nitrite and nitrate in scrubber liquors by manipulating the molarity of the eluent.   The development of measurement techniques for calcium and magnesium using the cation analytical column may also prove fruitful in scrubber studies.

Another aspect of the ion chromatograph of particular interest to our laboratory is its conduciveness to automation for the direct, semicontinuous monitoring of flue gas desulfurization processes. Because of its ease of use and relative compactness, it also lends itself to utilization in mobile field laboratories. Although much work still must be done on methodology, ion chromatography has already proved itself to be a useful tool in the field of stationary source and control device evaluation studies.

REFERENCES

1. Shawnee Test Facility Laboratory Procedures Manual, TVA (March 1976).
2. Steiber, R. and R. Statnick. "Simultaneous Determination of Sulfite and Sulfate by Ion Chromatography," *Ion Chromatog. Newsletter* (March 1977).
3. Mulik, J., R. Puckett, D. Williams and E. Sawicki. "Analysis of Nitrate and Sulfate in Ambient Aerosols," *Anal. Letters* 9(7):653-663 (1976).
4. Manual of Methods for Chemical Analysis of Water and Waste, EPA 625-/6-74-003 (1974), p. 277.
5. Goksoyr, H. and K. Ross. "The Determination of Sulfur Trioxide in Flue Gases," *J. Inst. Fuel* 35:177 (1962).

DISCUSSION

What standardization procedure do you use for sulfite?

The sulfite standard we use is sodium sulfite. We run an immediate ion chromatographic analysis and have the instrument calibrated for sulfite and sulfate.

Is the sulfate response in the sulfite standard due to sulfate in the actual sample?

Yes. For the sulfate we have found, we have done a set of analyses because we have to make up a standard that would be preservable. The sulfate we find is that which is initially present in the starting sodium sulfite solution. We can maintain the ratio of sulfite to sulfate fixed for about four hours, after which we get an ever-increasing conversion of sulfite to sulfate. Therefore, I believe that it is an autocatalytic activity such that we have essentially no increase in the amount of sulfate in our standard solution during the first four hours. In the next hour, we double it; in the hour after that it goes up almost by a factor of four. In eight hours, we have totally oxidized our solution. We are maintaining these solutions in a 100-ml volumetric flask and stopper after each sample is taken.

Did you try preserving the standard for more than four hours?

No. We felt that a four-hour lifetime for our standard was adequate for us.

How quickly was the sulfite oxidized to sulfate?

The initial solution would have about 1% of the total sulfur showing up as sulfate. At the end of eight hours, 100% of the sulfur would be showing up as sulfate.

Why don't you use peak area instead of peak height?

We have found that with our system we do better with peak height than with peak area. The data tend to scatter more when we include peak area. It may be something abnormal with our system or our data reduction procedure. The sulfite-sulfate peaks really tail negative very badly. It is very much a matter of judgment, where you want to cut the tail.

What potential do you think the method has?

I think that there is a potential for having it become the analytical finish. Now I think we have to back up and remember what we are dealing with here. We are not dealing necessarily with both a sampling and an analytical activity. You're talking basically about an analytical finish to a sampling activity that has preceded it. Now in several cases, the sampling activity precludes the use of ion chromatography. For example, if one wishes to determine acid sulfates by using 80% IPA as the collection media, that obviously precludes the use of ion chromatography. There are some problems with the IPA.

Did you try to prevent or slow down the oxidation?

We tried deaeration, but that had no effect. Apparently, we deaerated it in the process of taking samples and letting it sit in a stoppered volumetric flask. We allowed sufficient oxygen to reenter the system to permit the oxidation. We didn't look at the pH dependency of it. We started off with a basically acid solution because of the sulfite addition and did not monitor it. I think there have been a tremendous number of studies concerning sulfite oxidation to sulfate to support the flue gas desulfurization program, and they have shown that it is very strongly dependent upon the particular mix of anions and cations present in the solution, the dissolved oxygen content, and the mix of cations. That is where a lot of the pH dependency starts. The basic argument that most of those people put forward is that oxidation occurs by the bisulfite species and not the sulfite species.

13

ORGANIC ELEMENTAL MICROANALYSIS
BY ION CHROMATOGRAPHY

J. F. Colaruotolo

Hooker Chemicals and Plastics Corp.
Research Center, Grand Island Complex
Niagara Falls, New York 14302

ABSTRACT

Organic molecules containing chlorine, bromine, phosphorus and sulfur were decomposed by oxygen flask and the combustion products collected in 10 ml of distilled water containing 3 drops of 30% H_2O_2. After standing, this solution was diluted to 100 ml, and a sample was injected into a Dionex Model 10 Ion Chromatograph, which uses conductance as a detection system. Calibration curves constructed from the sodium salts of Cl^-, Br^-, PO_4^{-3} and SO_4^{-2} were linear over a wide range.

Determination of Cl, Br, P and S in organic molecules by this method resulted in the following accuracies and precisions for peak height and peak area, respectively: Cl accuracy $\pm$ 0.27, $\pm$ 0.42, precision $\pm$ 0.12, $\pm$ 0.25; Br accuracy $\pm$ 0.68, $\pm$ 0.58, precision $\pm$ 0.65, $\pm$ 0.48; P accuracy $\pm$ 0.39, $\pm$ 0.15, precision $\pm$ 0.10, $\pm$ 0.15; S accuracy $\pm$ 0.59, $\pm$ 0.25, precision $\pm$ 0.30, $\pm$ 0.25.

Molecules containing both Cl and Br were analyzed and the method found accurate to $\pm$ 0.5 for Cl and $\pm$ 1.5 for Br. In each case, total analysis time was less than one hour.

Two standards were introduced for use in standardizing the ion chromatograph for multi-element analysis. The potential exists for fast, convenient, simultaneous analysis of up to five heteroatoms.

INTRODUCTION

During the past several decades, there has been considerable effort expended in developing and improving methods of organic microanalysis. For any particular element one wishes to analyze, the literature offers a vast number of alternative methods. For those not familiar with the literature of organic microanalysis, the works of Ingram,[1] Siggia[2] and Olson[3] offer a good overview of the subject.

The advent of the Schoniger or oxygen flask technique[4] for the decomposition of organic compounds has greatly simplified the preparation of these materials for the analysis of their heteroatom content. Once the sample has been decomposed, any number of variations of several analytical techniques may be used to quantitate the element of interest. The reader should consult the literature for specific details; suffice it to say that the following techniques have been used to quantitate the element of interest after decomposition: gravimetry, titrimetry, colorimetry, X-ray fluorimetry, polarography, voltammetry and coulometry. Of these techniques, titrimetric methods are the most popular because of their relative ease of procedure, inexpensive apparatus and good accuracy.

Ion chromatography[5] has been shown to be an important new addition to this group of elemental microanalysis techniques.[6-8] All the above-mentioned techniques can suffer from interferences from other elements, which may be released during the decomposition step. For example, in the determination of chlorine and bromine in the same sample, the inherent chemical similarities of the two halogens makes it very difficult to obtain accurate results because the two elements interfere with each other. Cations cause severe interferences in the titrimetric determination of sulfur. Phosphorus will interfere with the gravimetric determination of sulfur and must be removed before the sulfur is precipitated. There are many imaginative analytical schemes in the literature, which have been developed to cope with interferences in elemental microanalysis. The ability of ion chromatography to separate ionic materials makes it an extremely powerful tool when applied to the problem of simultaneous multi-element analysis. It offers the possibility of very fast, accurate, simultaneous multi-element analysis without the necessary lengthy chemical preparations and conversions using the other techniques.

EXPERIMENTAL TECHNIQUE

Sample Preparation

Figure 13.1 shows a Schoniger flask, sample wrapper both before and after folding, and a combustion taking place. Several different absorption

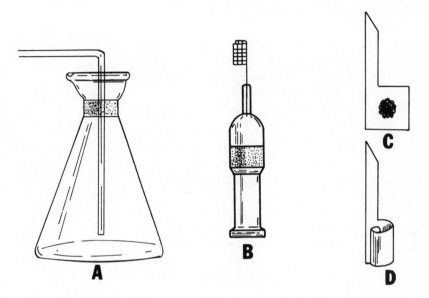

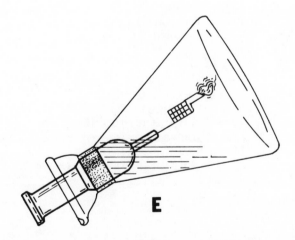

Figure 13.1 **A.** Schoniger or oxygen flask with oxygen inlet tube;
B. Flask stopper with platinum sample holder;
C. Ashless sample wrapping paper containing sample before folding;
D. Ashless sample wrapping paper containing sample after folding;
E. Sample combustion in flask containing absorption solution.

solutions have been found useful. Table 13.1 is a list of the more commonly used absorption solutions used in conjunction with the element of interest. In the case of halogen analysis, care must be taken to insure that the liberated halogen is reduced to the halide ion necessary for quantitation by ion chromatography. Therefore, the absorption solutions used for halides, especially bromine and iodine, contain reducing agents. For determinations of phosphorus and sulfur, oxidizing absorption solutions are necessary to convert them to phosphate and sulfate.

Table 13.1

Element	Absorption Solution	Ref.
Cl	10 ml H_2O	2
	10 ml 0.1 N NaOH	3
	10 ml H_2O + 3 drops 30% H_2O_2	9
	10 ml 0.1 N NaOH + 3 drops 30% H_2O_2	2
	2 ml 2 N KOH +10 ml H_2O + 3 drops 30% H_2O_2	10
	3 drops 30% H_2O_2 + 10 ml $CO_3^=/HCO_3^-$ eluent	7,8
Br, I	10 ml 2% hydrazine sulfate	10
Br	10 ml H_2O + 3 drops H_2O_2	6
Br, I	10 ml 2 N KOH + 10 ml 2% hydrazine sulfate	7
Br, I	Alkaline H_2O_2	3,4
Br, I	Alkaline bisulfite	3,11
Br, I	10 ml 0.01 N NaOH + 3 mg $NaBH_4$	12
	0.1% hydrazine in H_2O	8
F	10 ml H_2O	1
	0.1 N NaOH	2
P	5 ml bromine water + 5 ml 1 N NaOH + 5-10 ml H_2O	-
	10 ml H_2O + 3 drops 30% H_2O_2	6
	10 ml 1 to 2 N HNO_3	-
	3 drops 30% H_2O_2 + 10 ml $CO_3^=/HCO_3^-$ eluent	8
S	dilute NH_4OH + 3 drops 30% H_2O_2	2
	10 ml H_2O + 3 drops 30% H_2O_2	6,8,13
	3 drops 30% H_2O_2 + 10 ml $CO_3^=/HCO_3^-$ eluent	8

The absorption solutions used in conjunction with ion chromatography quantitation should not contain anionic materials that could cause interferences. Therefore, of those solutions listed in Table 13.1, those containing strong acids or bromine water are not desirable. Colaruotolo and Eddy have used 10 ml of water and 3 drops of 30% H_2O_2[6] exclusively for determination of Cl, Br, P and S, with good results. Smith, McMurtrie and Galbraith[7] and Smith, McMurtrie and Scheidl[8] have used 3% H_2O_2/ 0.003 M $NaHCO_3$/0.0024 M Na_2CO_3 for Cl and S and hydrazine in aqueous base for Br and I. The hydrazine does not interfere with the analysis since, in the cation suppressor column, it is converted to an amine salt and remains on the column.

After combustion and absorption are complete, the sample can either be injected directly[7] or made up to a known volume and a portion of this dilution injected.[6]

Calibration of the Ion Chromatograph

Several ways are available for calibration of the instrument. Standard solutions can be prepared from the sodium salts of the anions of interest. The stock solution from which these standards are prepared by series dilution must be standardized by an accurate analytical method. The chloride and bromide stock solution is standardized by potentiometric titration with $AgNO_3$, the phosphate by precipitation as its quinoline molybdate complex, and sulfate by precipitation as barium sulfate. An alternative method of calibration is to combust several samples of a standard organic material of known heteroatom content using a range of sample weights. Figures 13.2 to 13.9 are calibration curves of Cl, Br, PO_4 and SO_4 produced from standard solutions of their sodium salts using both peak height and peak area. Figures 13.10 and 13.11 are calibration curves produced from the combustion of standard organic materials. The calibration curves are linear to less than 1 mg/l levels. We have found it most convenient to work in the 1 to 100 mg/l level. Deviation from linearity due to the fundamental conductance of the ions can occur at about 50 mg/l in a pure solution and is also dependent on what other species are present. Carbonic acid generated from the mixed carbonate eluent, being only weakly ionized, has relatively little effect on the anions of interest. In practice, depending on the ion injected and its separation volume, deviation from linearity occurs at about 100 mg/l.

Analysis of Samples

Table 13.2 summarizes the instrumental parameters that have been used both by our laboratory and by others. Figure 13.12 shows typical outputs we have obtained for combusted compounds containing chlorine, bromine, phosphorus and sulfur, using the same conditions in Table 13.2 for all anions.

Small blanks due to residual chloride in the sample wrapper and distilled water had to be subtracted from peak heights or areas due to

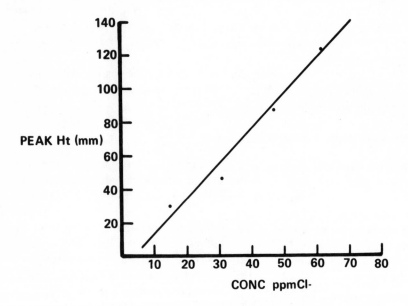

Figure 13.2   Chlorine calibration curve, peak height vs concentration.

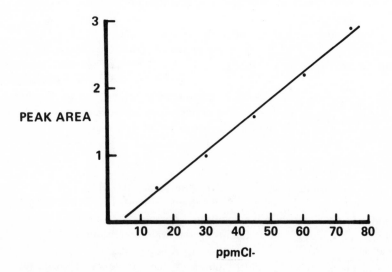

Figure 13.3   Chlorine calibration curve, peak area vs concentration.

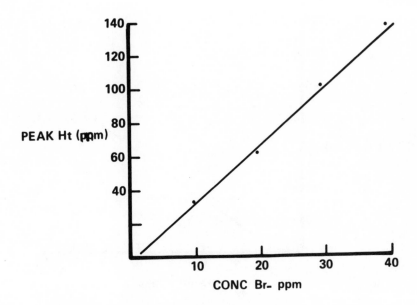

Figure 13.4 Bromine calibration curve, peak height vs concentration.

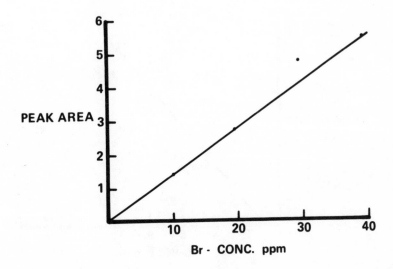

Figure 13.5 Bromine calibration curve, peak area vs concentration.

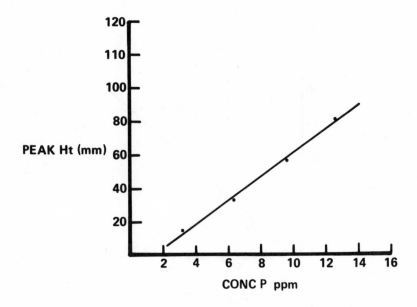

Figure 13.6 Phosphorus calibration curve, peak height vs concentration.

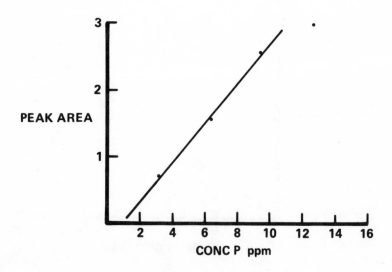

Figure 13.7 Phosphorus calibration curve, peak area vs concentration.

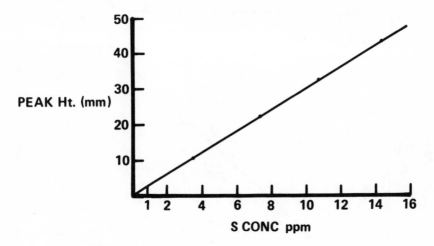

Figure 13.8 Sulfur calibration curve, peak height vs concentration.

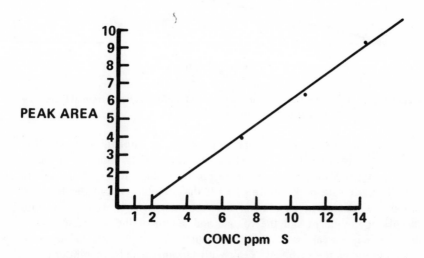

Figure 13.9 Sulfur calibration curve, peak area vs concentration.

chloride from samples. The carbonate generated from the carbonaceous portion of the organic sample plus the sample wrapper caused a small carbonate peak over and above the background carbonate eluent. This interfered slightly with the bromide peak but was reproducible and simply subtracted out as a blank. The carbonate blank was reproducible even though different organic molecules have more or less carbon, and sample wrappers do not always weigh exactly the same. This is because

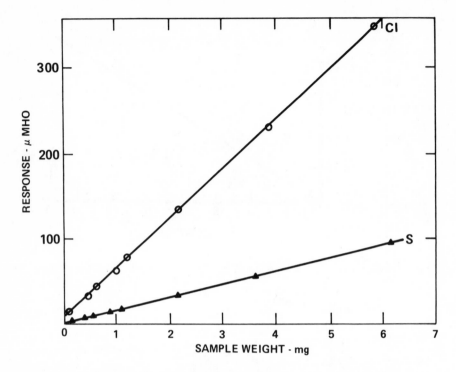

Figure 13.10   Calibration curves for chlorine and sulfur prepared from the combustion of S-benzyl thiuronium chloride.[7]

carbonate emerges from the suppressor column as carbonic acid, which is only feebly dissociated and, therefore, the conductivity detector is relatively insensitive to changes in the carbonate blank. It should be noted that the carbonate peak interfered with the bromide peak under the conditions we used. The position of the carbonate peak is affected by both the pH of the eluent and the volume and length of the suppressor column, as it is a weak base. Smith, McMurtrie and Galbraith[7] demonstrated the mobility of the carbonate peak with chromatographic conditions.

RESULTS AND DISCUSSION

Table 13.3 summarizes the results obtained in our laboratory using the first set of chromatographic conditions in Table 13.2 for all the analyses, and using standards prepared from the sodium salts of the anions. Meta-chlorobenzoic, o-bromobenzoic and triphenylphosphate were National Bureau of Standards standard reference materials 2144, 2142 and 147, respectively. The other compounds were purified by four recrystallizations

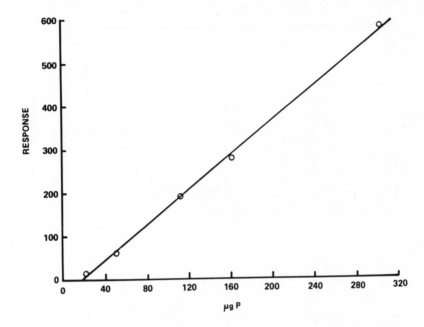

Figure 13.11   Calibration curve for phosphorus prepared by combusting triphenyl-phosphine samples.[8]

Table 13.2 Chromatographic Conditions—Dionex Model 10 Ion Chromatograph

1.[6]   Eluent: 0.003 M NaHCO$_3$/0.0024 M Na$_2$CO$_3$
   Flow rate: 1.5 ml/min.
   Sample loop: 0.1 ml
   Separator column: 3 mm ID x 500 mm
   Separator resin: Dionex low-capacity anion exchange resin DA-X5-0.565 surface
             agglomerated to Dionex surface-sulfonated polystyrene divinyl
             benzene copolymer DCS-X2-55
   Suppressor column: 9 mm ID x 500 mm
   Suppressor resin: Dionex high-capacity cation exchange resin DC-X12-55
   Detection: conductivity

2.[8]   Eluent: 0.0030 M NaHCO$_3$/0.0024 M Na$_2$CO$_3$
   Flow rate: 138 ml/hr for chlorine; 230 ml/hr for sulfur
   Sample loop: 200 μl
   Separator column: 3 mm ID x 500 mm
   Separator resin: Dionex low-capacity anion exchange resin
   Suppressor column: 6 mm ID x 250 mm
   Suppressor resin: Dionex high-capacity cation exchange resin
   Detection: conductivity

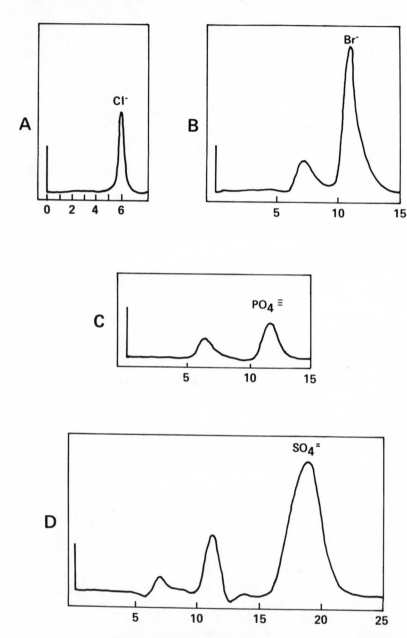

Figure 13.12 **A.** Chloride output from the combustion of *o*-Cl-benzoic acid;
B. Bromide output from the combustion of *o*-Br-benzoic acid;
C. Phosphate from the combustion of triphenylphosphate;
D. Sulfate output from the combustion of sulfanilamide.

Table 13.3 Elemental Analysis Using IC

Compound	Weight (mg)	Peak Height	Peak Area
		Cl Found (%)	
m-Cl-Benzoic Acid	7.08	23.02	23.92
(22.64% Cl Theory)	6.41	22.93	22.91
	6.27	22.96	22.59
	8.50	22.71	23.06
	8.19	23.20	23.32
o-Cl-Acetanilide	7.66	20.89	21.15
(20.90% Cl Theory)	6.40	20.78	21.25
	5.97	20.60	20.94
p-Br-chlorobenzene	8.01	17.64	18.35
(18.52% Cl Theory)	7.05	17.27	18.84
4'-Cl-2 Br-acetophenone	7.21	15.07	15.64
(15.18% Cl Theory)	8.56	15.09	16.31
		Br Found (%)	
o-Br-Benzoic Acid	7.05	39.57	38.58
(39.75% Br Theory)	8.11	39.95	40.07
	5.94	40.57	40.07
	7.01	37.30	39.80
	6.86	39.87	41.25
5-Br-Salicylic Acid	4.00	37.37	36.50
(36.82% Br)	7.10	36.20	37.25
	7.85	37.32	37.32
p-Br-chlorobenzene	8.01	39.39	40.29
(41.73% Br Theory)	7.05	40.16	43.06
4'-Cl-2 Br-acetophenone	7.21	34.83	35.88
(34.22% Br Theory)	8.56	33.32	36.99
		P Found (%)	
Triphenylphosphate	8.41	9.69	9.69
(9.4% P Theory)	10.22	9.98	9.58
Triphenylphosphine	7.12	11.23	11.52
(11.81% P Theory)	8.54	11.22	-
Sulfanilamide	9.29	19.01	18.41
(18.58% S)	9.26	19.62	19.10
	7.46	18.89	18.63

from ethanol. For single-element analysis, the agreement between the observed value and the theoretical value was excellent. From these data the absolute accuracy and standard deviation were calculated for each element. Table 13.4 summarizes the results of these calculations for the determinations using both peak height and area. Tables 13.5 and 13.6 summarize the results of Smith, McMurtrie and Galbraith.[7] They compared results of chlorine and sulfur determinations by ion chromatography with those obtained by argentimetric titration and barium precipitation, respectively. The agreement between ion chromatography and the more classical methods is good. Of particular note in Table 13.6 are the results obtained for low levels of sulfur. The ion chromatographic method produced excellent low-level sulfur results with sample weights of 29 to 150 mg, while to carry out a successful barium precipitation determination, 3 to 5 g of sample were required.

Table 13.4   Absolute Accuracy and Standard Deviation for
Cl, Br, P and S Determinations

Element	Absolute Accuracy (%)		Standard Deviation (%)	
	Peak Height	Peak Area	Peak Height	Peak Area
Chlorine	± 0.27	± 0.42	0.17	0.42
Bromine	± 0.68	± 0.58	1.09	0.82
Phosphorus	± 0.39	± 0.15	0.15	0.10
Sulfur	± 0.59	± 0.25	0.32	0.29

Smith, McMurtrie and Scheidl[8] have described the application of ion chromatography to the following four problems in heteroatom analysis of organic compounds: (1) the determination of trace sulfur in organic compounds; (2) the analysis of compounds containing mixed halogens; (3) the determination of phosphorus in organic compounds; and (4) the rapid determination of trace nitrogen in aqueous samples after the nitrogen species are converted to ammonia by digestion and total N is determined by ion chromatography as NH_4^+ after distillation. Their comparison of sulfur determinations by ion chromatography and barium gravimetric analysis, for sulfur content of 4 to 12%, showed the methods to agree on the average to within ± 0.1%. For trace-level sulfur between 50 and 300 ppm sulfur, the agreement between ion chromatography and barium turbidimetry averaged ± 14 ppm. Ion chromatography gave consistently better reproducibility than the barium technique.

Table 13.5 Comparison of Chlorine Percentages by Ion Chromatography (IC)
and Argentimetric Titration (AT)[7]

Sample	Weight for IC Analysis (mg)	Chlorine Percent	
		IC	AT
56	2.272	40.47	39.90
57	10.000	2.28	2.39
58	10.984	2.35	2.43
59	4.095	15.44	15.44
60	2.140	1.65	1.71
70	0.532	49.20	47.38
71	0.526	34.22	31.60
72	1.185	11.64	13.20
73	1.056	7.04	6.7
74	16.921	0.54	0.25
	16.201	0.52	0.63
75	16.804	0.41	0.45

Table 13.6 Comparison of Sulfur Percentages by Ion Chromatography (IC) to
Known Values (KV) and Barium Precipitation Techniques (BPT)[7]

Sample	Weight for IC Analysis (mg)	Sulfur Percent		
		IC	BPT	KV
NBS Coal A	22.420	0.546	-	0.546
NBS Coal B	10.812	1.97	-	2.02
NBS Coal C	8.279	3.09	-	3.02
Sulfamic Acid	7.469	32.95	-	33.03
	0.501	32.94	-	33.03
80	3.988	7.29	7.25	-
81	10.900	1.52	1.46	-
82	1.560	12.59	11.52	-
	0.522	12.49	-	-
84	43.6	0.0150	0.0240	-
85	96	0.0326	0.0297	-
86	159	0.0258	0.0296	-
	29	0.0239	0.0331	-
87	110	0.0259	0.0288	-

They compared analysis of chlorine in the presence of bromine by ion
chromatography with chlorine results obtained by argentimetric titration.
Agreement between the two methods averaged ± 0.15%. Before the argentimetric titration could be carried out, the bromine was removed by boiling

the sample in HNO_3. No pretreatment was necessary for ion chromatography because Cl^- and Br^- were completely resolved.

Analysis of phosphorus after combustion was also examined using ion chromatography. Triphenyl phosphine (11.81% P) standard was burned in Schoniger flasks with sample weights from 0.1 to 3 mg. Using a 3% H_2O_2 in eluent scrubber, a linear calibration curve was observed. Their work on phosphorus is continuing.

Lastly, results by ion chromatography for trace-digestible nitrogen after Kjeldahl digestion were compared with those obtained by the more conventional Nessler analysis. Agreement between the two methods averaged ± 22 ppm for samples containing from 100 to 700 ppm N.

Consideration of the results discussed above gives some perspective in which to view ion chromatography relative to the other available methods of elemental analysis. For the case of single–element analysis, ion chromatography cannot replace the more conventional methods, but simply takes its place with them as an additional tool for analysts to use. However, for multi-element, simultaneous analysis, the limited data available suggest that ion chromatography will eventually completely replace the conventional methods.

The last two entries in Table 13.3 under Cl Found and Br Found summarize the results obtained in our laboratory for the analysis of bromine and chlorine in two different mixed halogen compounds. Figure 13.13 shows the separation obtained for Cl^- and Br^- under the conditions of the first entry in Table 13.2. For chlorine, the accuracy was ± 0.58 and ± 0.52 for peak height and peak area, respectively, while for bromine, it was ± 1.30 and ± 1.79. The advantage of this analysis is that the chromatographic time for elution of both elements was only 10.5 minutes. The available methods for mixed halogen analysis require several hours of analysis time per sample so that for many applications, where high accuracy is not required, the ion chromatographic method has a clear advantage.

Some unique standards have been synthesized at Hoffman LaRoche[8] (Figure 13.14) for use in standardizing the ion chromatograph for multi-element analysis. Compound 1 is a multiple halogen standard and compound 2 is a multiple halogen plus sulfur standard. Figure 13.15 is a chromatogram of compound 1 mixed with sulfamic acid using 0.1% N_2H_4 as the absorber solution. The potential exists for fast, convenient, simultaneous analysis of up to five heteroatoms.

CONCLUSIONS

Ion chromatography has been shown to be a useful alternative to the more conventional methods of heteroatom analysis of organic materials

Figure 13.13 Separation of chloride and bromide from the combustion of $4'$-Cl-2-Br-acetophenone.

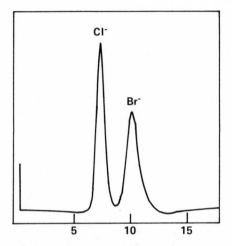

ION	–%
F	3.83
Cl	7.14
Br	16.09
I	25.56

Figure 13.14 Standards for multi-elemental simultaneous analysis.[8]

following oxygen flask decomposition. The feasibility of rapid simultaneous multi-elemental microanalysis has been demonstrated by several workers. Continued effort on simultaneous multi-elemental analysis will produce very fast and accurate methods that ultimately will be automated.

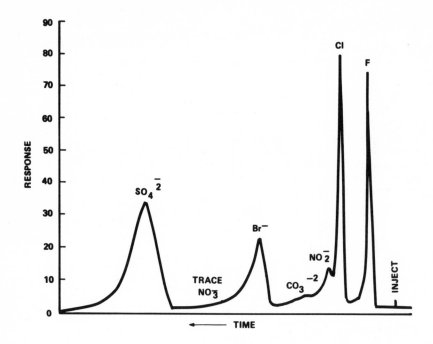

Figure 13.15   Chromatogram of compound 1 mixed with sulfamic acid.[8]

REFERENCES

1.   Ingram, C. *Methods of Organic Elemental Microanalysis* (New York: Reinhold Publishing Corp., 1962).
2.   Siggia, S. *Survey of Analytical Chemistry* (New York: McGraw-Hill Book Co., 1968), p. 38.
3.   Olson, E. C. in *Treatise on Analytical Chemistry,* I. M. Kolthoff and P. J. Elving, Eds. (New York: Wiley-Interscience, 1971), p. 1.
4.   Schoniger, W. *Mikrochim. Acta* 869 (1956).
5.   Small, H., T. S. Stevens and W. C. Bauman. *Anal. Chem.* 47:1801 (1975).
6.   Colaruotolo, J. F. and R. S. Eddy. *Anal. Chem.* 49:824 (1977).
7.   Smith, F. Jr., A. McMurtrie and H. Galbraith. American Chemical Society National Meeting, San Francisco, California, August 1976.
8.   Smith, F., Jr., A. McMurtrie and F. Scheidl. 28th Annual Pittsburgh Conference, Cleveland, Ohio, February 1977.
9.   Olson, E. C. and A. F. Krivis. *Microchem. J.* 4:181 (1960).
10.   ASTM. "Standard Test Method E442-74 for Chlorine, Bromine or Iodine in Organic Compounds by Oxygen Flask Combustion," *Annual Book of ASTM Standards,* Part 30 (1976), p. 842.
11.   Steyermark, A. and M. B. Faulkner. *J. Assoc. Office Agr. Chem.* 35:291 (1952).

12. Celon, E. *Mikrochim. Acta* 592 (1969).
13. ASTM. "Standard Test Method E443-74 for Sulfur in Organic Compounds by Oxygen Flask Combustion," *Annual Book of ASTM Standards*, Part 30 (1976), p. 847.

DISCUSSION

Have you determined any phosphates with ion chromatography?

We do not see any polymeric phosphate species in the samples of tripolyphosphate and pyrophosphate we have run ion chromatographically. We don't see them in their combustions. And as far as low values, phosphate has been our easiest compound to analyze. We don't have a problem with water or peroxide.

What do you use as an absorbing solution for the halogens?

We use hydrazine and KOH as an absorbant solution for the mixed halogens or for bromine by itself. Any bromine in the solution that is not bromide will be reduced to bromine by hydrazine. Again, we haven't had too much trouble with our bromine determinations, but we haven't been doing that many mixed halogens. We could run into problems in the future.

14

QUANTITATIVE DETERMINATION OF INORGANIC SALTS IN CERTIFIABLE COLOR ADDITIVES

D. D. Fratz

Division of Color Technology
U.S. Food and Drug Administration
Washington, D. C. 20204

ABSTRACT

In the certification of color additives for use in foods, drugs and cosmetics, it is useful to quantitatively determine any inorganic salts present. Most commonly found are sulfates and chlorides that have formerly been determined through separate potentiometric titrations. The titration for chloride actually determines total halide content. Phosphate and carbonate give a positive interference in the sulfate titration procedure. A new procedure has been developed, using a Dionex Ion Chromatograph equipped with an anion exchange column, to produce quantitative determinations of numerous inorganic ions in certifiable color additives in a single analysis. A small anion exchange precolumn was used to trap the dye after injection. The procedure has been used to analyze 27 soluble dyes for fluoride, chloride, nitrite, phosphate, bromide, nitrate and sulfate ions. A method using a shorter analytical column was developed to determine iodide. The salts were determined using peak height response, and calculated as percent sodium salt.

INTRODUCTION

In the certification of color additives for use in food, drugs and cosmetics, it is presently necessary to perform routine determinations of chloride, sulfate and sometimes iodide. Analyses for chloride and sulfate formerly have been accomplished through separate potentiometric titrations.[1,2] The titration for chloride, however, actually determines total halide content, and phosphate and carbonate give a positive interference in the sulfate titration procedure. Iodide formerly has been determined by a lengthy elution chromatographic procedure.[3] In addition, the determination of other inorganic salts during color certification, should they be present, would also be of value in evaluating the integrity of the manufacturing process.

For these reasons, evaluation of the Dionex Ion Chromatograph was initiated to determine its possible usefulness in determining a broad spectrum of inorganic salts in color additives. The determination of numerous inorganic salts in aqueous solution has previously been accomplished using the ion chromatograph.[4]

A procedure was developed which traps the color and its subsidiary colors and intermediates on a precolumn, allowing the inorganic salt ions to pass through for anion exchange separation. The method can determine fluoride, chloride, nitrite, phosphate, bromide, nitrate and sulfate in a single determination in 27 water-soluble certifiable color additives. A separate method, using a shorter analytical column, was developed for the determination of iodide.

The water-soluble color additives analyzed in this method are mostly grouped in two major classes—the sulfonated azo dyes and the halogenated fluorescein dyes. The azo dyes can be grouped into three subclasses. The first is formed by a coupling with a β-naphthol derivative, and includes FD&C Red No. 40, FD&C Yellow No. 6 and D&C Orange No. 4 (Figure 14.1). The second group is formed by a coupling with an α-naphthol derivative, and includes FD&C Red No. 4 and D&C Red No. 33 (Figure 14.2). The third group is formed by a coupling with pyrazolone, and includes FD&C Yellow No. 5 and Orange B (Figure 14.3).

The two subclasses of fluorescein dyes are very similar, one being the acid form and the other the sodium salt. The acid form includes D&C Orange No. 5, D&C Orange No. 10, D&C Red No. 21, D&C Red No. 27 and D&C Yellow No. 7 (Figure 14.4). The sodium salt form includes FD&C Red No. 3, D&C Orange No. 11, D&C Red No. 22, D&C Red No. 28 and D&C Yellow No. 8 (Figure 14.5).

Two other classes of certified colors are the triphenylmethanes, represented by FD&C Blue No. 1, D&C Blue No. 4 and FD&C Green No. 3, and the indigoids, represented by FD&C Blue No. 2 (Figure 14.6). Seven other water-soluble certifiable colors belong to various other classes and subclasses (Figure 14.7).

Only one of the water-soluble certifiable color additives was found unsuitable for this procedure—D&C Red No. 19, a xanthene derivative

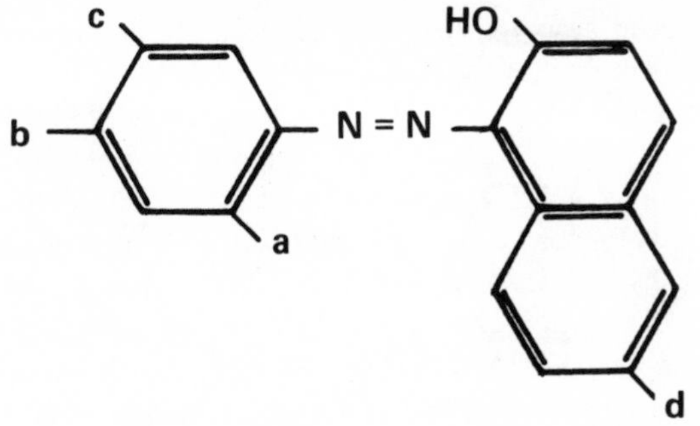

	a	b	c	d
FD&C RED NO. 40	OCH3	SO3Na	CH3	SO3Na
FD&C YELLOW NO. 6	H	SO3Na	H	SO3Na
D&C ORANGE NO. 4	H	SO3Na	H	H

Figure 14.1 Structures of some 1-phenylazo-2-naphthol dyes.

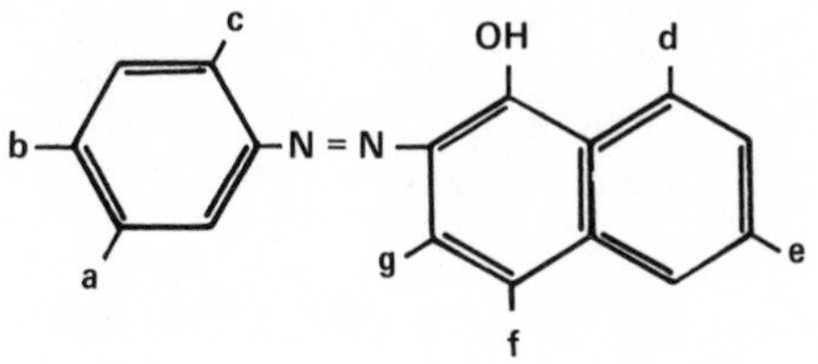

	a	b	c	d	e	f	g
FD&C RED NO. 4	SO3Na	CH3	CH3	H	H	SO3Na	H
D&C RED NO. 33	H	H	H	NH2	SO3Na	H	SO3Na

Figure 14.2 Structures of some 2-phenylazo-1-naphthol dyes.

(Figure 14.8). The precolumn would not hold back the dye, which sticks to the suppressor column.

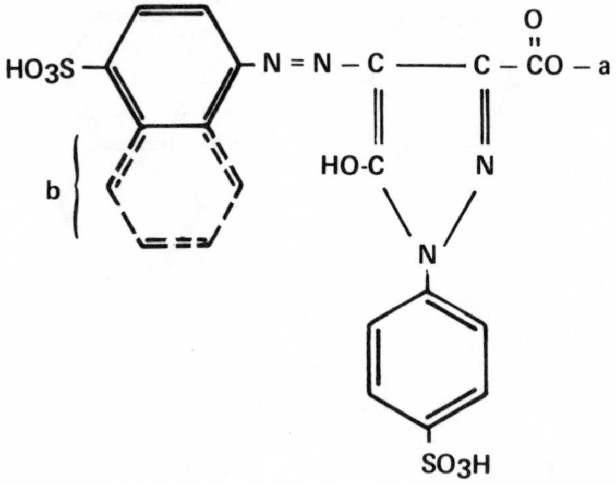

	a	b
FD&C YELLOW NO. 5	H	WITHOUT
ORANGE B	C_2H_5	WITH

Figure 14.3   Structures of some 1-phenyl-3-carboxy-4-arylazo-2-pyrazolin-5-one dyes.

It was found that the widths of the peaks generated were stable enough in all the species determined to base the calculations on peak height alone without any calculation of peak area.   Therefore, the determination is quite simple to calculate and no integrator is needed.

METHOD

Apparatus

1. Dionex Ion Chromatograph.   Model 10 or 14 (Dionex Corporation, 1228 Titan Way, Sunnyvale, California 94086).
2. Analytical column.   3- x 500-mm Chromix DCS-X2-55 substrate resin with Chromex DA X5-0.376 microspherical agglomerating resin.
3. Precolumn and iodide analytical column.   3- x 150-mm Chromex DCS-X2-55 substrate resin with Chromex DA X5-0.376 microspherical agglomerating resin.
4. Suppressor column.   6- x 250-mm Chromex DC-X12-55.
5. Chart recorder.   Any model suitable for ion chromatographic use.

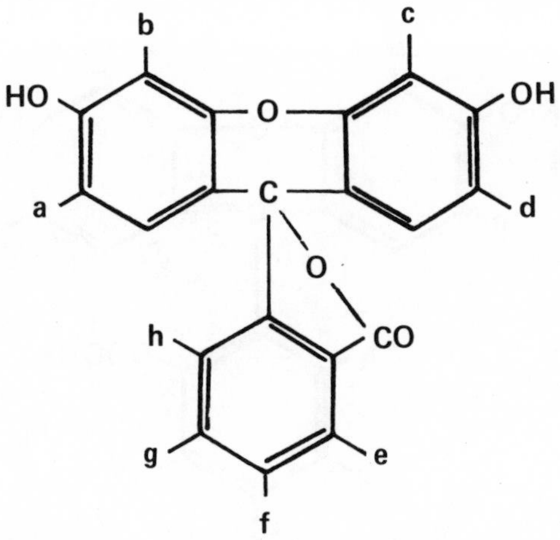

	a	b	c	d	efgh
D&C ORANGE NO. 5	H	Br	Br	H	H
D&C ORANGE NO. 10	H	I	I	H	H
D&C RED NO. 21	Br	Br	Br	Br	H
D&C RED NO. 27	Br	Br	Br	Br	Cl
D&C YELLOW NO. 7	H	H	H	H	H

Figure 14.4 Structures of some halogenated fluorescein dyes—acid form.

Reagents

1. Eluent (E2). 0.003 N sodium bicarbonate/0.0024 N sodium carbonate. (Add 1.00 g sodium bicarbonate and 1.00 g sodium carbonate to 4 liters distilled water and stir to dissolve.)
2. Precolumn regenerate. 0.1 N HCl/50% acetone. (Carefully pour 83 ml conc. HCl into 400 ml distilled water in a 1-liter mixing graduated cylinder, bring to 500 ml with distilled water, and bring to 1 liter with acetone.)
3. Suppressor regenerate (RI). 0.1 N sulfuric acid.
4. Analytical column reconditioner (EI). 0.1 N sodium carbonate.
5. External standard solutions. Stock solution: Dissolve 1.00 g NaCl and 500 mg each of sodium fluoride, sodium nitrite, trisodium phosphate, sodium bromide, sodium nitrate and sodium sulfate in 1 liter

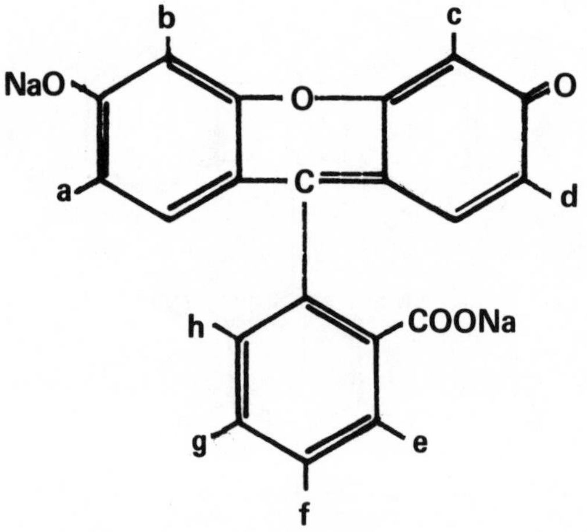

	a	b	c	d	efgh
FD&C RED NO. 3	I	I	I	I	H
D&C ORANGE NO. 11	H	I	I	H	H
D&C RED NO. 22	Br	Br	Br	Br	H
D&C RED NO. 28	Br	Br	Br	Br	Cl
D&C YELLOW NO. 8	H	H	H	H	H

Figure 14.5   Structures of some halogenated fluorescein dyes—sodium salts.

eluent. Dilute stock solution 1:1000, 1:250, 1:100 and 1:50 with eluent to prepare low, medium, high and very high standard solutions, respectively. This equals a range of 0.5 to 10.0% NaCl, and 0.25 to 5.0% for the other salts, calculated on a w/w sodium salt/ dye basis.

6. Iodide standard solutions. Stock solution: Dissolve 100 mg sodium iodide in 1 liter eluent. Dilute stock solution 1:100, 1:50, 1:20 and 1:10 to get 0.5, 1.0, 2.5 and 5.0% standards, respectively, calculated on a w/w sodium iodide/dye bases.

7. Sample solutions. 0.2 mg color/ml eluent. Add 20.0 mg color sample to 100 ml eluent and dissolve by stirring.

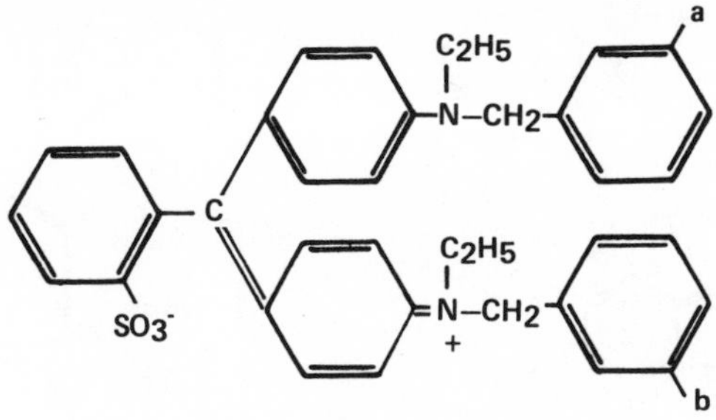

	a&b
FD&C BLUE NO. 1	SO3Na
D&C BLUE NO. 4	SO3NH4

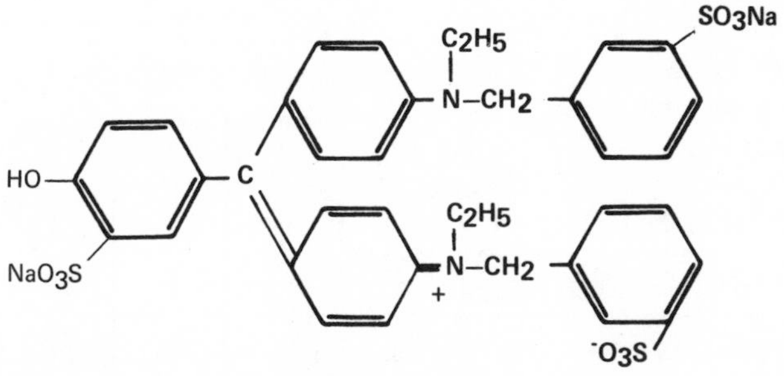

FD&C GREEN NO. 3

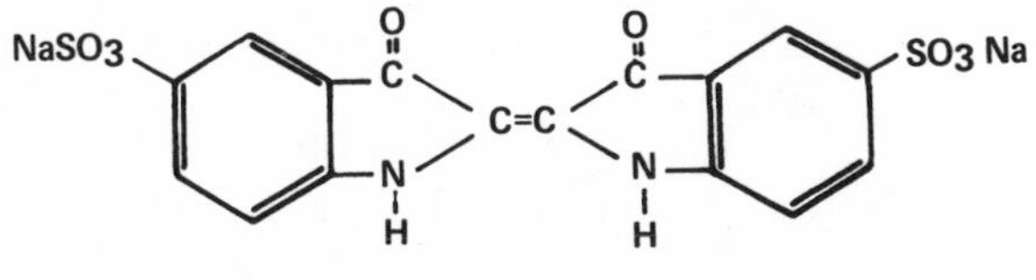

Figure 14.6 Structures of some triphenylmethane and indigoid dyes.

Procedure

Set pump speed at 30% (124 ml/hr). Offset specific conductance to 0 when background is stable. Standardize by using the loop injector to inject 100 μl of each standard, with a scale range of 10 μmho, switching to 1 μmho after chloride has eluted. If the results are linear ± 3%, the instrument is standardized. The ions will elute in the following order: fluoride, chloride, nitrite, phosphate, bromide, nitrate and sulfate.

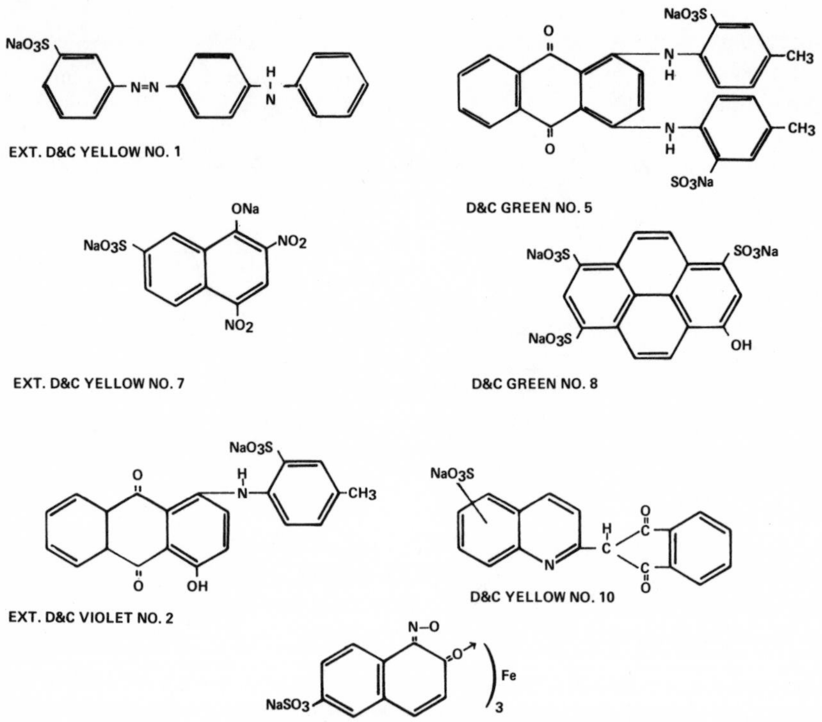
EXT. D&C YELLOW NO. 1
D&C GREEN NO. 5
EXT. D&C YELLOW NO. 7
D&C GREEN NO. 8
EXT. D&C VIOLET NO. 2
D&C YELLOW NO. 10

Figure 14.7 Seven other certifiable water-soluble color additives.

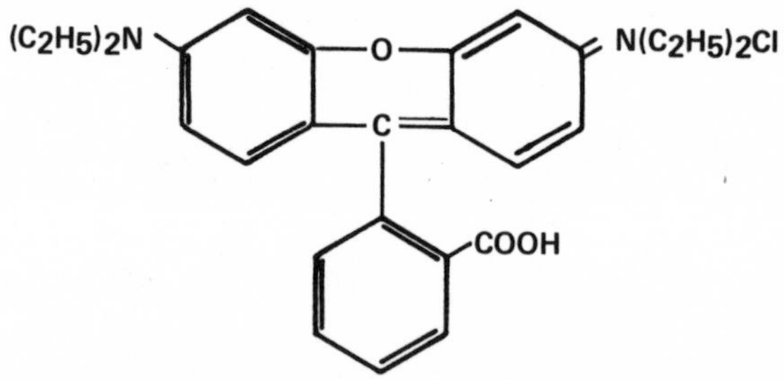
(C2H5)2N
N(C2H5)2Cl
COOH

Figure 14.8 Structure of D&C Red No. 19.

Inject 100 μl sample solution (20 μg color additive) with the scale range set at 10 μmho. Once again, after chloride elutes (*ca.* 3 min), change scale range to 1 μmho. After sulfate elutes, change scale range back to 10 μmho and inject the next sample.

Suppressor column regeneration and precolumn regeneration must be done approximately every eight hours. Suppressor column is regenerated using the automatic regeneration cycle set for 15 minutes of regenerate solution followed by a 30-minute distilled water rinse. The precolumn is regenerated by taking the column off-line and pulling about 15 ml precolumn regenerate solution through the column with a syringe, to remove the color additives, and then forcing through approximately 15 ml distilled water to remove the regenerate solution.

The analytical column should be reconditioned once a month, or whenever retention begins to be lost, to remove any ions retained by the column. With the suppressor column off line, flow through reconditioning solution (EI) for about 30 minutes, followed by eluent (E2).

In the method above, the elution time for iodide is too long for the procedure to be used for its determination. If, however, the 500-mm column previously used is eliminated, and pump speed set at 50%, determination of iodide can be achieved, using only the precolumn.

Calculations

The amounts are calculated by measuring the peak height (specific conductance) of each species in the sample and comparing to the peak height of the nearest standard.

$$\% \text{ Na salt} = \frac{\text{Specific conductance of sample peak x \% Na salt in standard}}{\text{Specific conductance of standard peak}}$$

RESULTS AND DISCUSSION

The chromatograms obtained by the standardization procedure show baseline separations on all the ions except bromide and nitrate. The resolution is, however, sufficient to allow the peak height response to be unaffectedly determined (Figure 14.9). Iodide elutes long after sulfate by this method, as a very low, broad peak detectable only as a very minor baseline drift, and does not interfere with later determinations. When the 500-mm analytical column is eliminated in the iodide determination, the elution time for iodide is lowered, allowing the iodide to be determined (Figure 14.10).

Figures 14.11 a, b and c show the standard curves obtained by plotting peak height response vs percent sodium salt for each of the ionic species to be determined. In each case, the results show sufficient linearity to allow determinations calculated using the peak height response without necessitating the calculation of peak area. The actual peak heights obtained will vary from column to column and day to day on the same column.

The elution times and relative sensitivities given in Table 14.1 are also subject to variation. Columns will sometimes begin to lose retention with age, causing the resolution and elution times to decrease, and raising the sensitivity as the peaks become narrower. Loss of retention without

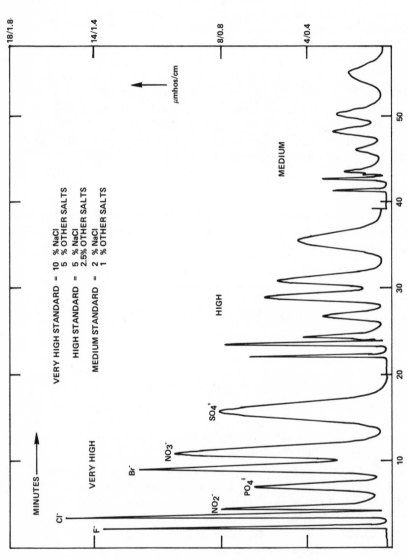

Figure 14.9 Ion chromatograms obtained by the standardization procedure for fluoride, chloride, nitrite, phosphate, bromide, nitrate and sulfate for three concentrations.

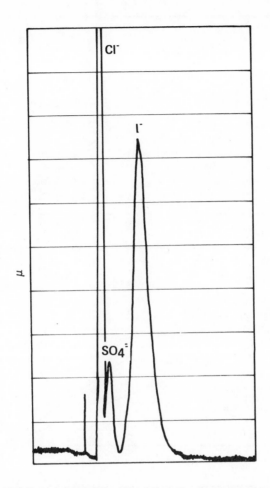

Figure 14.10   Ion chromatogram of a FD&C Red No. 3 sample with 5% iodide added.

increase in sensitivity is sometimes caused by channeling, and can be partially corrected by running the analytical column at higher flow rates for short periods of time.

Table 14.2 shows the results of the analysis of commercial samples of each of 27 water-soluble color additives, both as received for certification and with small amounts of each salt added. None of the dyes was observed to interfere in the determination of any of the salts.

Samples of FD&C Red No. 3 and D&C Orange No. 10, both iodinated fluoresceins, were analyzed for iodide, and none was found.

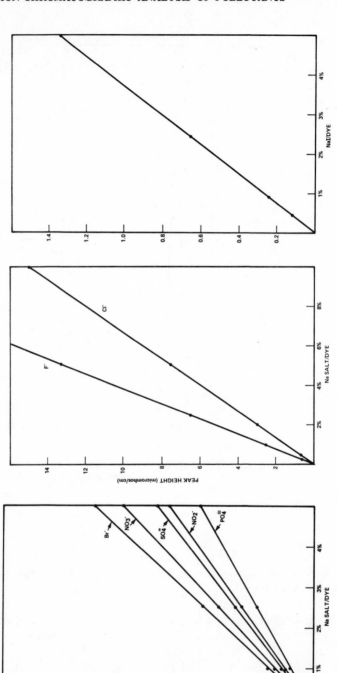

Figure 14.11 Standard curves obtained by plotting peak height response vs percent sodium salt for (a) fluoride and chloride, (b) bromide, nitrate, sulfate, nitrite and phosphate and (c) iodide.

Table 14.1   Typical Elution Times and Relative Sensitivities of Various Ions

	F^-	Cl^-	NO_2^-	$PO_4^{\equiv}$	Br^-	NO_3^-	$SO_4^{=}$	I^- [b]
Elution Time (min.)	2.1	3.5	4.4	6.7	9.0	11.0	15.5	8.5
Relative Peak Height Sensitivity[a]	1.8	1.0	0.10	0.08	0.15	0.13	0.11	0.017

[a] In specific conductance peak height per weight sodium salt relative to chloride.
[b] By iodide method.

Samples with iodide added at various levels, however, showed good recoveries, as seen in Table 14.3.

The 1 N hydrochloric acid/50% acetone precolumn regeneration solution must not be used within the instrument. The metal fittings and air lines on the chromatograph are very susceptible to corrosion from HCl fumes.

ACKNOWLEDGMENTS

The author would like to thank Courtney Anderson and William Rich for their extensive help in the development of this method.

REFERENCES

1. Graichen, C. and J. Bailey. "Automatic Potentiometric Titration of Sodium Chloride in Certifiable Water-Soluble Colors," *J. Assoc. Off. Anal. Chem.* 57:356 (1974).
2. Bailey, J. and C. Graichen. "Automatic Potentiometric Titration of Sodium Sulfate in Certifiable Water-Soluble Colors," *J. Assoc. Off. Anal. Chem.* 57:353 (1974).
3. Link, W. "Intermediates in Food, Drug and Cosmetic Colors," *J. Assoc. Off. Anal. Chem.* 44:43 (1961).
4. Small, H., T. Stevens and W. Bauman. "Novel Ion Exchange Chromatographic Method Using Conductimetric Determination," *Anal. Chem.* 47 (11):1801 (1975).

DISCUSSION

Do organic compounds interfere in the ion chromatographic procedure?

No. The larger organic species, which include intermediates such as 2-naphthylamine and B-naphthol, are all held at the precolumn. They do not come through at the concentrations we find.

Why is it necessary to analyze food dyes for anions?

In food dyes, the specifications are for the anions. There are specifications for chlorides, sulfate and iodide for each of the food dyes.

Table 14.2 Determinations of Salts in Commercial Color Additive Samples and the Same Samples Spiked with 5% Cl and 2.5% Other Salts

	Weight/Weight Percent Sodium Salt						
	F^-	Cl^-	NO_2^-	$PO_4^{\equiv}$	Br^-	NO_3^-	$SO_4^=$
FD&C Blue No. 1	0.05	0.10	nf	nf	nf	nf	0.25
spiked sample	2.5	5.2	2.6	2.4	2.5	2.5	2.8
FD&C Blue No. 2	nf	1.9	nf	0.15	nf	nf	0.90
spiked sample	2.4	7.1	2.6	2.7	2.5	2.5	3.5
D&C Blue No. 4	nf	nf	nf	nf	nf	nf	0.60
spiked sample	2.5	5.1	2.5	2.5	2.6	2.4	3.2
FD&C Green No. 3	nf	1.3	nf	nf	nf	nf	0.55
spiked sample	2.4	6.1	2.5	2.4	·2.5	2.6	3.0
D&C Green No. 5	nf	3.3	nf	nf	nf	nf	0.55
spiked sample	2.4	8.4	2.7	2.4	2.6	2.7	3.0
D&C Green No. 8	nf	19.9	nf	0.10	nf	nf	0.50
spiked sample	2.5	25.3	2.6	2.5	2.6	2.7	2.9
Ext. D&C Green No. 1	nf	2.0	nf	nf	nf	nf	0.15
spiked sample	2.5	7.1	2.6	2.4	2.5	2.7	2.7
Orange B	nf	3.0	nf	nf	nf	nf	0.15
spiked sample	2.5	3.1	2.4	2.5	2.5	2.6	2.6
D&C Orange No. 4	nf	3.0	nf	nf	nf	nf	0.05
spiked sample	2.5	8.2	2.5	2.5	2.5	2.5	2.5
D&C Orange No. 5	0.20	1.1	nf	nf	0.15	nf	0.90
spiked sample	2.7	6.0	2.6	2.6	2.7	2.5	3.4
D&C Orange No. 10	nf	0.80	nf	nf	nf	nf	0.15
spiked sample	2.6	5.9	2.5	2.4	2.4	2.5	2.7
FD&C Red No. 3	nf	0.70	nf	nf	nf	nf	0.10
spiked sample	2.5	5.9	2.6	2.5	2.6	2.5	2.6
FD&C Red No. 4	nf	2.2	nf	0.10	nf	nf	0.15
spiked sample	2.5	7.2	2.5	2.6	2.5	2.5	2.6
FD&C Red No. 40	nf	4.3	nf	nf	nf	nf	0.10
spiked sample	2.5	9.4	2.6	2.6	2.5	2.5	2.7
D&C Red No. 21	nf	0.55	nf	nf	nf	nf	0.20
spiked sample	2.6	5.6	2.6	2.6	2.5	2.5	2.80
D&C Red No. 22	nf	0.25	nf	nf	0.45	nf	0.10
spiked sample	2.5	2.8	2.7	2.6	3.0	2.5	2.7
D&C Red No. 27	nf	0.30	nf	nf	0.30	nf	0.15
spiked sample	2.5	5.3	2.6	2.5	2.9	2.5	2.7
D&C Red No. 28	0.05	0.20	nf	nf	2.5	nf	0.35
spiked sample	2.5	5.3	2.6	2.5	4.9	2.6	2.8
D&C Red No. 33	nf	6.6	nf	0.15	nf	nf	0.65
spiked sample	2.5	11.9	2.5	2.6	2.5	2.5	3.2
Ext. D&C Violet No. 2	0.10	5.7	nf	0.40	nf	0.10	0.20
spiked sample	2.6	10.9	2.6	3.0	2.6	2.6	2.7

Table 14.2, continued

	Weight/Weight Percent Sodium Salt						
	F^-	Cl^-	NO_2^-	$PO_4^{\equiv}$	Br^-	NO_3^-	$SO_4^=$
FD&C Yellow No. 5	nf	2.1	nf	nf	nf	nf	0.15
spiked sample	2.5	1.0	2.7	2.5	2.5	2.5	2.6
FD&C Yellow No. 6	nf	3.2	nf	nf	nf	nf	0.10
spiked sample	2.5	8.2	2.5	2.6	2.5	2.5	2.6
D&C Yellow No. 7	nf	0.20	nf	nf	nf	nf	0.05
spiked sample	2.5	5.2	2.6	2.5	2.5	2.5	2.5
D&C Yellow No. 8	nf	0.10	nf	0.80	nf	nf	0.25
spiked sample	2.4	4.9	2.4	3.4	2.7	2.6	2.8
D&C Yellow No. 10	nf	5.4	nf	nf	nf	nf	0.10
spiked sample	2.4	10.6	2.5	2.5	2.5	2.5	2.6
Ext. D&C Yellow No. 1	nf	0.40	nf	nf	nf	nf	0.10
spiked sample	2.4	5.4	2.6	2.5	2.5	2.5	2.7
Ext. D&C Yellow No. 7	nf	2.7	nf	nf	nf	nf	0.15
spiked sample	2.4	7.7	2.7	2.4	2.6	2.6	2.7

Table 14.3 Recoveries of Iodide in FD&C Red No. 3 and D&C Orange No. 10

	Unspiked	+ 5% NaI	+ 2.0% NaI	+ 0.5% NaI
FD&C Red No. 3	nf	5.2%	2.1%	0.4%
D&C Orange No. 10	nf	5.4%	2.3%	0.6%

15

ION CHROMATOGRAPHY: AN
ANALYTICAL PERSPECTIVE

W. E. Rich, J. A. Tillotson and R. C. Chang
Dionex Corporation
Sunnyvale, California 94086

ABSTRACT

This chapter discusses the general advantages of the chromatographic method of analysis. It places ion chromatography (IC) in an analytical perspective relative to other types of chromatography, with emphasis on the classes of compounds and applications problems uniquely analyzable by IC. Future developments including automation, new detectors and higher performance are discussed relative to their projected impact on solving problems involving ions in solution.

INTRODUCTION

During the past 20 years, chromatography (both gas and liquid) has been evolving toward high-speed separations coupled with continuous effluent monitoring by detectors that provide "real time" analysis of the eluted species. The chromatographic method has proved itself to be a powerful analytical tool for the qualitative and quantitative analysis of a vast number of organic and inorganic compounds. The advantages of the chromatographic method of analysis are well known: speed, where multi-element analysis per sample is often accomplished in less than an hour; resolution, which can always be compromised for speed and gives the researcher the ability to resolve interferences; sensitivity, in most cases unusually high compared with classical methods of analysis; and, finally, automation, which is straightforward and adds another dimension to problem-solving efficiency where routine operations are preferred.

185

To place ion chromatography in its proper perspective among other chromatographic methods, consider the compound matrix consisting of low-molecular-weight ($\sim < 2000$) organic and inorganic compounds described in Figure 15.1. The matrix ranges in one direction from nonvolatile to volatile compounds and in the other, nonpolar to polar to ionic to compounds highly acidic or basic in solution; that is, having a low $pK_{A \text{ or } B}$. Dividing modern chromatography into gas chromatography (GC) and high-pressure liquid chromatography (HPLC) and assigning each a region of optimum analysis capability, it is clear that GC is most effective in analyzing volatile compounds that extend in polarity over the entire region, while LC covers the entire polarity region for nonvolatile compounds. There are, of course, overlapping regions. Further, subdivide LC into liquid-liquid chromatography (LLC), liquid-solid chromatography (LSC) and ion exchange chromatography (IEC) and assign LLC and LSC the nonpolar to ionic region and IEC the ionic to low pK region. HPLC has been very successful in utilizing various forms of packings such as reverse phase and normal phase packing materials to analyze nonpolar, polar and certain ionic compounds, most having chromophores. High-performance ion exchange packings, either pellicular or microporous, are excellent for ionic separations but are limited by and large to chromophoric species due to the lack of a sensitive nonchromophoric detector. Also, pellicular packings tend to chemically decompose at extreme pH ranges sometimes required for elution of highly basic or acid compounds. It is intrinsic to normal IEC that sample ions absorb in the UV or visible regions of the spectrum; otherwise, sensitive, universal detection is almost impossible. In certain cases, such as amino acid analysis, nonchromophoric ions are combined with other chemicals (ninhydrin) to form chromophores, but there are few other examples of this type.

Due to the lability of most ion exchange HPLC packings, the swelling problems associated with microporous polystyrene HPLC packings and the necessity for the ions of interest to have chromophores, the low $pK_{A \text{ or } B}$ region of this compound matrix could not, in general, be analyzed. Certain highly acidic or basic organic compounds having strong chromophores can be analyzed by Paired-Ion HPLC. But this exception is similar to amino acid analysis where special chemicals must be added for the analysis to work.

Ion chromatography,[1] with its special separating resin and suppressor column and conductimetric detector, allows direct routine analysis of compounds with low $pK_{A \text{ or } B}$, with or without chromophores. The classes of compounds that have been analyzed by IC to date include inorganic anions, organic acids (mostly aliphatic), alkali metals, alkali earth metals, ammonia, alkylamines and quaternary ammonium compounds. It

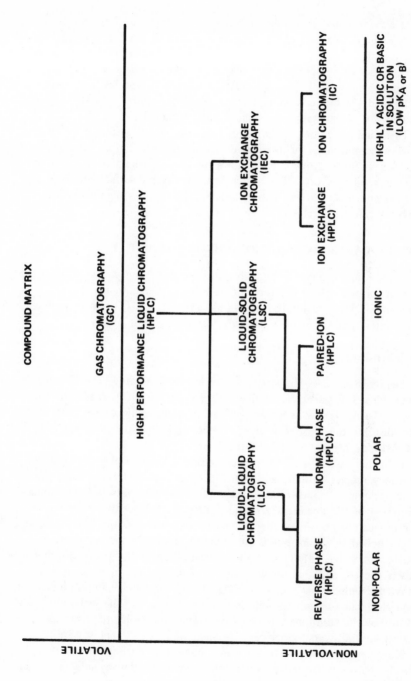

Figure 15.1 Ion chromatography in an analytical perspective relative to other types of chromatography.

must also be noted that volatile gases such as HCl and SO_2, which may be difficult to analyze by GC, can be analyzed by IC if converted to a highly basic or acidic, water-soluble form. It is important to note that these classes of compounds are analyzed to the exclusion of (1) weakly dissociated species, $pK_{A \text{ or } B} > {\sim}7$, (2) amphoteric species (*i.e.*, amino acids), and (3) species that undergo "negative" chemistry on the suppressor column (*i.e.*, precipitation of some transition metals.)

These unique consequences result from the use of a suppressor column. Thus, uncomplicated highly resolved chromatograms can be obtained from complex matrices. Figure 15.2 shows the chromatogram of water from secondary oil recovery. This water contains organic surfactants, high salt content and many different ionic species. However, a highly resolved chromatogram is obtained.

UNIQUE CAPABILITIES OF ION CHROMATOGRAPHY

Ion chromatography is uniquely powerful in solving problems involving ions in solution where (1) very high sensitivity is required, (2) multi-element detection facilitates the sample analysis, and (3) matrix problems cause interferences for other analysis techniques.

High Sensitivity

Figure 15.3 shows the separation of low-level nitrate and sulfate in rainwater. This particular sample contains 4.1 ppm NO_3^- and 3.6 ppm $SO_4^=$. Levels as low as a few ppb can be easily achieved for these ions and illustrate the sensitivity of IC. Other techniques, such as wet chemical methods, are not reliable at these levels.

Industrial boilers used for process steam and electrical power generation may require periodic and costly overhaul caused by corrosion and scale build-up in boiler tubes. These problems are controlled by chemical treatment of the demineralized and deaerated boiler feedwater and by continuous blow-down of the boilers; *i.e.*, a small fraction of the feedwater is not boiled to steam, but is continuously drained from the boiler as blow-down water. Monitoring the concentration of the treatment chemicals is needed, and this can be done by analyzing the blow-down water. Ethylenediaminetetraacetic acid (EDTA) and orthophosphate are added to sequester scale-forming calcium and magnesium ions. The orthophosphate added has the additional effect of passivating the boiler tube surface against general overall corrosion. Sodium sulfite treatment scavenges residual oxygen from the boiler water, preventing pitting corrosion. Finally, corrosion caused by hydrogen ion is controlled by adding sodium

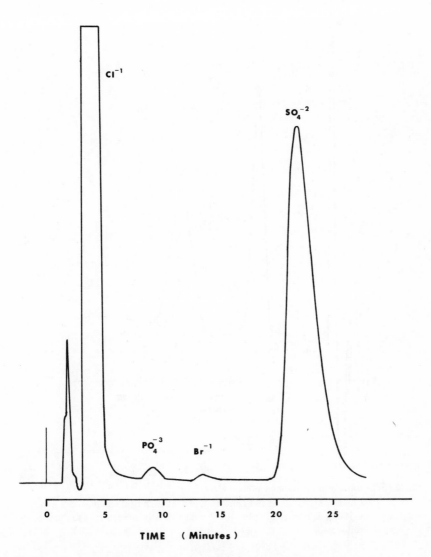

Figure 15.2   Ion chromatogram of a water sample from secondary oil recovery.
Analytical column: 2.8 mm x 750 mm packed with Dionex low-
capacity anion agglomerated resin; suppressor column: 6 mm x
250 mm packed with Dionex high-capacity cation exchange resin;
eluent: 0.003 M NaHCO$_3$/0.0024 $\overline{M}$ Na$_2$CO$_3$; flow rate: 161 ml/hr;
conductimetric detector: 10 μmho/cm, full-scale; sample size: 100 μl.

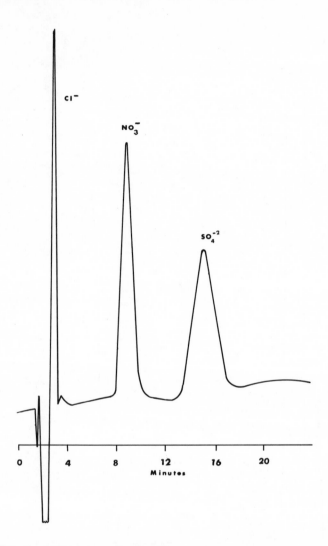

Figure 15.3   Ion chromatogram of a rain water.
Conditions: Same as in Figure 15.2 except analytical column: 2.8 mm
x 500 mm; flow rate: 138 ml/hr; conductimetric detector: 1 μmho/cm;
full-scale.

hydroxide. Stevens[2] has used an automated IC at the Dow Chemical
Company to monitor these waters directly. Figure 15.4 shows a typical
chromatogram obtained from boiler blow-down water by IC. Quantitation
of these ions can be achieved by area or peak height measurement at low
ppm levels.

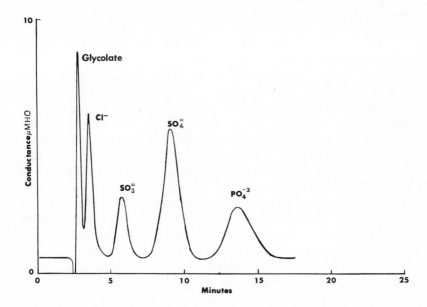

Figure 15.4   Ion chromatogram of boiler blow-down water.

Conditions: Same as in Figure 15.2 except analytical column: 2.8 mm x 1000 mm; suppressor column: 2.8 mm x 300 mm; eluent: 0.005 M Na_2CO_3/0.004 M NaOH; flow rate: 115 ml/hr; sample size: 500 μl.

Multi-element Detection

Multi-element detection facilitates the analysis of ground waters and drinking water as illustrated in Table 15.1. Here, five ions are directly analyzed in each sample of water collected from six different locations around the San Francisco Bay Area.

Table 15.1 Anion Content of Ground Water Samples

	F^- (ppm)	Cl^- (ppm)	NO_3^- (ppm)	$SO_4^=$ (ppm)
Sunnyvale Tap Water	0.065	47	1.61	31.32
Nancy Zellhoefer's Well	0.45	82	22.3	12.2
Crystal Springs Reservoir	0.027	3.6	0	4.2
San Francisquito Creek[a]	0.22	161	4.5	529
Stevens Creek Reservoir	0.164	44	0	117.7
Lake Vasona[a]	0.26	28.5	0.86	213

[a]Trace Br also present.

Matrix Problems

The analysis of anions in brine (2 to 25% NaCl) and caustic (50% NaOH) solutions is extremely difficult due to the high chloride and hydroxide levels, respectively. Figure 15.5 shows the direct analysis of $SO_4^=$ in 25% brine following a 500-fold dilution. Figure 15.6 shows the analysis of Cl^- and ClO^-, ClO_3^- and $SO_4^=$ in 50% caustic. Table 15.2 summarizes typical analyses performed by IC in a pulp and paper industry. It shows various types of samples with particular ions of interest and their respective concentration ranges.[3] These are examples of complex matrices in which IC offers direct analyses where other methods have interferences or are cumbersome and time-consuming.

These are just a few examples of the unique capabilities of IC and how they can be utilized in practical, analytical problem-solving. Other classes of compounds, such as organic acids and amines, have special analysis problems of their own. IC offers the same unique capabilities here as well. In addition, thermally labile, nonchromophoric organic compounds can many times be better analyzed by IC than GC or HPLC.

FUTURE DEVELOPMENTS OF ION CHROMATOGRAPHY

The evolution of IC will be somewhat different from that of GC and HPLC. Performance capabilities for both GC and HPLC were developed through improved instrumental versatility and specifications, column packings and detectors, before automation was added. The reverse will be true for IC. The reason for this reversal is not completely clear, but it concerns the ease of applying IC to a wide range of different applications, the class selectivity of its detection system, and a current trend toward automation of instrumentation.

Automation will occur in two phases: to automate (1) the current laboratory IC systems, and (2) a more rugged system for quality control applications.

Performance improvements could be made in two major areas: resolution and new detectors. Higher resolution involves a higher pressure, more precise pumping and valving system, higher efficiency column packing materials, reduction of dead volume in the system including the suppressor column and more selectivity control, such as temperature and gradient programming and ion exchange resins with a wide variety of selectivities. New detectors such as potentiometric, coulometric and UV/ visible types could either replace or be added to the current suppressor column/conductivity scheme. New detectors would expand the "analytical window" of IC to include more of the ion exchange chromatographic

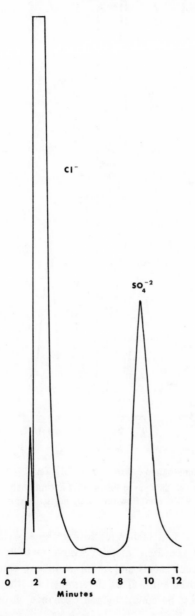

Figure 15.5 Ion chromatogram of 25% brine.

Conditions: Same as in Figure 15.2 except analytical column:
2.8 mm x 500 mm; flow rate: 184 ml/hr; conductimetric detector:
1 μmho/cm, full-scale; sample: 500:1 dilution.

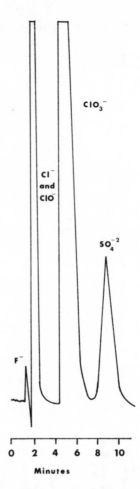

Figure 15.6   Ion chromatogram of 50% caustic solution.
Conditions: Same as in Figure 15.2 except analytical column:
2.8 mm x 500 mm; flow rate: 207 ml/hr; sample: 100:1 dilution.

region, more than would higher resolution alone. This would make possible the analysis of transition metals, organic metallic compounds and a wider range of organic acids and amines. In any case, the evolution of performance improvement will be determined by the immediacy of problems that might best be solved by ion exchange chromatography development.

Table 15.2 Typical Analysis Performed by IC in the Pulp and Paper Industry

Sample Type	For	Concentration Range
Hatchery Water	NO_3^--N	20 - 60 ppb
	$PO_4^\equiv$-P	5 - 20 ppb
Paper Mill Effluent	$SO_4^=$-S	10 - 100 ppm
Biopond Water	NO_3^--N	10 - 50 ppb
	$PO_4^\equiv$-P	100 - 175 ppb
KCl Soil Extract	$SO_4^=$	100 - 300 ppm
Acetate Soil Extract	$SO_4^=$	10 - 100 ppm
Ammonium Fluoride Soil Extract	$PO_4^\equiv$	1 - 10 ppm
Sulfuric Acid Soil Extract	$PO_4^\equiv$	10 - 50 ppm
Bicarbonate Soil Extract	$SO_4^=$	1 - 5 ppm
Pulping Liquid (Green)	S^{-2}	very high
	$SO_4^=$	very high

CONCLUSIONS

IC has enabled the direct chromatographic analysis of a class of compounds heretofore difficult to analyze by existing methods. It has paved the way for new methods development in ion exchange chromatography. With the future developments outlined in this chapter, the class of compounds analyzable by IC will certainly increase. Automation of IC will quickly expand its use in quality control and process monitoring applications. However, the growth of IC as well as ion exchange chromatography will be determined by the success of researchers in applying this technique to real problems involving ions in solution.

REFERENCES

1. Small, H., T. S. Stevens and W. C. Bauman. "Novel Ion Exchange Chromatographic Method Using Conductimetric Detection," *Anal. Chem.* 47:1801 (1975).
2. Stevens, T. S. Paper presented at the Federation of Analytical Chemistry and Spectroscopy Societies, Philadelphia, Pennsylvania, November 1976.
3. Barnes, E. and R. C. Chang. Unpublished results, 1976.

DISCUSSION

What do you foresee as future improvements in ion chromatography?

I believe that as far as technology will go, I see improvement, first of all, in the efficiency of the columns. I see vast improvements coming in commercially available ion exchange pellicular-type packings, in increases in stability and performance so that they could be used in suppressor columns, and conductivity meters to do higher-performance ion chromatography. I think that day is very close. I see as the next phase, though, improving selectivity control—that means changing the actual chemical surface of the ion exchange resin by changing the basicity of the ion exchange site and grossly changing the selectivity of one ionic form over the other one. Again, I would like to point out that these changes probably won't be derived out of basic research as much as out of the need to solve problems.

Isn't the cost of analysis actually higher than what you pictured in the figure?

We didn't know what to choose, so we put in the equation and decided it looked good. We just used the depreciation on the instrument. You're right. The cost should be higher than that.